7 Divine Weapons of Mass Destruction In the

King's Armory

Paul K. Adom

For permission requests and any other enquiries write to the address below.

Paul K. Adom
PO Box 13703, Durham NC 27709
https://pauladom.com

ISBN: 978-1-7358748-0-7 (Paperback Edition)

ISBN: 978-1-7358748-1-4 (eBook Edition)

ISBN: 978-1-7358748-2-1 (Hardcover Edition)

My deepest appreciation to these designers on https://www.fiverr.com/

Illustrations by Hanson Suhendra (onesketchman)

Cover Design by Germancreative

Book Layout by Marko Markovic (Marmarko78)

Printed in United States of America
for Worldwide Distribution

First Edition

DEDICATION

I dedicate this work to my Heavenly Father, JEHOVAH
The Man of War, The LORD Strong and Mighty,
The LORD Mighty in Battle! Thank you, Father,
for making me a battle-ax of your kingdom.

I dedicate this book to Jesus Christ my Lord, my Savior, my King, my War General, and Commander in Chief! Thank you, Lord, for teaching my hands to war and my fingers to battle.

I dedicate this book to the Holy Spirit, the Spirit of Wisdom and Revelation who guided my hand throughout this project.

I dedicate this book to Heaven's End Time War General and God's Face to Face Prophet, my Spiritual Father Apostle David E Taylor, who taught me the lost message of the Kingdom.

I dedicate this book to my dear wife Pastor Judy Adom who has been a great encouragement and who helped get this book ready for publication.

I dedicate this book to my dear daughter, Anna, a powerful shaker that God will use mightily in the days to come.

And I dedicate this book to Kingdom Restoration Church For All Nations, the seal of my apostleship. You have been the firstfruits of this work!

Thank you, Constance Hunt, for the major boost you gave me to start this work!

I must also make special mention of Professors Edward and Eunice Akoto who helped edit the book. Your work on this project was priceless.

Table Of Contents

PREFACE

THE LORD GAVE ME THIS revelation of divine weapons of war in 2009 and 2010 when I was doing a two-year teaching series on Hebrews 11 in Durham NC. In the Summer of 2010, I taught this revelation in a powerful Spiritual Warfare workshop where I expounded on several divine weapons of war. For the past ten years, I have walked out this revelation and now the Lord is having me share with the Body of Christ some of these specialized weapons of war in scripture and how to harness them for the warfare of the End Times.

As I was planning this war manual, I sketched on paper the cover page. I had a cross in the middle of the front cover with the other weapons in four sections around the cross. Then one Sunday, one of my elders shared a dream she had with me. In this dream, God showed her a book I had written. It was a small book and the front cover had a cross on it with other weapons. She then handed me a drawing she had made of what she saw. It was just like mine! And now here it is. Therefore I know that this is a God-ordained book that will bless you immeasurably. This book will teach you certain spiritual warfare tactics and strategies that will help you win battles!

This book makes use of pictures to drive home the main truths. "A word fitly spoken is like apples of gold in pictures of silver." (Pro 25:11) The ancient world communicated with pictures as you can see in some of their inscriptions and writings on walls of old caves. The Hebrew Old Testament Bible was originally written in the ancient pictographic Hebrew script comprising 32 symbols. The word "God" or "El" in ancient Hebrew is two pictures – an ox head representing strength, and a shepherd's staff symbolizing authority. (See https://www.ancient-hebrew.org/index.html). I believe it is time for the church to recover this ancient art of communication because God often talks in pictures and images.

This book is critical for the end times because a major spiritual conflict brewing over the last few generations is about to explode into a full-scale spiritual war between the kingdom of God and the Kingdom of Satan. God is therefore stirring up His end-time army for the greatest battle ever! Joel 2:2 describes this army as **"a great people and a strong; there hath not been ever the like, neither shall be any more after it, even to the years of many generations."** For many generations, there will be no war like the one we are about to experience. For this purpose, God is equipping His end-time army with high-level divine weapons of war like the ones you are about to learn in this book. It is time for you my dear friend to master these weapons!

Over the next few years, I will be sharing more of these spiritual weapons to equip you for the end-time battle. Now, may this book embolden and strengthen you to wage a good warfare. May you shake and turn your region upside down as you employ the 7 divine weapons of mass destruction revealed in this book, in Jesus' mighty name!

Paul Kwame Adom

pkadom@icloud.com
https://pauladom.com
https://kingdomrestoration.church
https://kingdominvasion.net

INTRODUCTION

DIVINE WEAPONS OF MASS DESTRUCTION

GOD HAS RESERVED FOR HIS people powerful weapons as the bible says, "**For the weapons of our warfare** are not carnal, but mighty through God to the pulling down of strong holds" (2 Cor 10:4). The Lord used the word "weapons" because there is more than one weapon. These weapons are "mighty through God" in that they are divine or spiritual weapons that God has forged to equip us to win spiritual battles being waged for the destiny of entire cities and nations of the world. During our walk with the Lord, He trains our hands to war and our fingers to fight with these spiritual weapons (Psa 144:1). I call these weapons divine weapons of mass destruction because with them you can destroy principalities and powers oppressing entire regions.

Weapons of mass destruction are not ordinary weapons. According to Encyclopedia Britannica, a Weapon of Mass Destruction (WMD) is a ***"weapon with the capacity to inflict death and destruction on such a massive scale and so indiscriminately that its very presence in the hands of a hostile power can be considered a grievous threat. Modern weapons of mass destruction are either nuclear, biological, or chemical weapons—frequently referred to collectively as NBC weapons."***

In much the same way, God has divine weapons of mass destruction (WMD) infinitely more powerful than the WMD that man has developed. These divine WMD destroy demonic infra-structures that Satan has established over entire regions. These are specialized weapons that can win whole regions for God. Each of the 7 weapons you are about to study in this book are like massive nuclear bombs that can destroy satanic strongmen ruling entire regions, setting free millions and billions of souls oppressed in their strongholds.

This book is therefore for believers who want to go into a region and take it over for the Lord. With these weapons, believers are going to end major spiritual wars that are being waged over cities and nations just like the atomic bombs America dropped over Hiroshima and Nagasaki ended the Second world war. With these divine WMD, believers will enter a city, cast out the ruling prince, and see the whole city gather around them just like Jesus did during His earthly ministry. "**And all the city was gathered together at the door**" (Mark 1:33). God is raising mighty warriors in this hour who will experience this!

You must therefore be willing to wage the kind of warfare that will break demonic princes oppressing entire regions. It is the violent who will do battle to possess their inheritance and the promises of God for their life (see Matthew 11:12). Jesus never won a single city until He first wrestled with the Devil in the wilderness for 40 days and 40 nights and cast him out saying "Get thee hence, Satan" (Matt 4:10). It is only then that we see angels come and minister to Jesus. And that is when His fame began to spread throughout the land as He preached the gospel of the Kingdom with mighty signs and wonders. We all want to see this happen, but who is willing to do battle, wage war, and combat satanic strongmen, and powers? You cannot just go into a city and walk off with the souls. You must first overcome and bind the principality or strongman who has established a hold on that city. Jesus said,

> *Or else how can one enter into a strong man's house, and spoil his goods,* ***except he first bind the strong man?*** *and then he will spoil his house* (Matthew 12:29).

You need specialized weapons that can destroy that strongman and his stronghold. That is what a divine weapon of mass destruction does. This kind of weapon overpowers the strongman, takes from him his armor, and levels his stronghold to the ground! When you are done with this book you will be equipped with 7 of these weapons. But first I need you to understand that God expects you to rule the stronghold of ***your*** life as a prelude and prerequisite to ruling your region. Because how can you rule your city when you do not rule your life?

> ***He that hath no rule over his own spirit is like a city that is broken down, and without walls*** (Pro 25:28)

You must dethrone the strongman of your life before you can dethrone the strongman ruling a city. You must first apply these weapons in your life before you use them to shake your region. This is why Matthew 7:5 says, "Thou hypocrite, **first cast out the beam out of thine own eye;** and then shalt thou see clearly to cast out the mote out of thy brother's eye." When you rule your life, you will *"see clearly"* or you will have great ease in taking over your region with these weapons! When you rule your life God will give you power and authority to rule your city and break the strongman oppressing your region! As you win your personal battles with these divine weapons, you will be empowered to tear down strongholds of darkness in your region! And so, in this book, you will learn how to wield these weapons in your own personal life and in your region. To rule your life, you must walk in your identity and office as a king.

YOU MUST WAGE WAR AS A KING

To benefit from this book, you must understand your office and mandate as a king. That is a major foundational point in this spiritual warfare manual. I want you to understand that you are a king with dominion and so when God created you, He crowned you "... with glory and honour" (Psa 8:5). That crown identifies you as a king! The bible says, "The LORD is King for ever and ever" (Psa 10:16). God is a King and He created you in His image and likeness to have dominion on earth as a King (See Gen 1:26). God said to our father Abraham in Genesis 17:6 "... kings shall come out of thee." As a seed of Abraham, you are one of the kings who came out Abraham's loins! Jesus redeemed you by His blood to make you a king and a priest!

> ***And hast made us unto our God kings** and priests: and **we shall reign on the earth*** (Revelation 5:10).

Revelation 1:6 also states, "**And hath made us kings** and priests unto God and his Father..." God made you a king to exercise dominion on earth. This is why this book is titled "7 Divine Weapons Of Mass Destruction In The ***KING'S*** armory." **It is Kings who put on armor and do battle!** Second Samuel 11:1 says, "... at the time when **kings go forth to battle...**" It is Kings who go forth to battle. God ordained you in your kingship to wage war against the kingdom of darkness. The Lord Jesus Christ gave an amazing revelation about this to Apostle David E. Taylor which he speaks about in his book "The Empire of God"

> **"He said to me, *"You have been trying to face a demonic kingdom with a priestly anointing."* This is exactly what is happening in the body of Christ.**

> **We have tried to face a demonic kingdom with a church anointing and it does not work... Demonic princes rule masses of people from the air! They are royal spirits, not church demons. You cannot break these demonic royal spirits with a church anointing because it takes a king to take out a king. It takes a kingdom to confront another kingdom! Only kings have power to deal with kings. Until we understand this, we will not be able to turn our cities over to the Lord."**
> *(The Empire of God Part 1 by David E Taylor; page 29)*

What an amazing revelation of the Kingdom! Only a king can take out another king! I believe that no one in the body of Christ teaches the Kingdom of God better than David E Taylor. His teachings on the Kingdom are so unprecedented because Jesus appeared to him face to face in many dreams and visions and gave him that revelation. In your warfare, you are contending with evil spirits who wear crowns and sit on thrones in heavenly places. You must therefore wage war against them out of your kingship identity and anointing, not just out of your priesthood. You have an anointing to fulfill your kingship assignment, just like David, Solomon, and Jehu were anointed to be kings in Israel. The Bible says,

---

> *"A king that sitteth in the throne of judgment scattereth away all evil with his eyes"* (Proverbs 20:8).

You must sit on the throne of judgment as a King to scatter the wicked. Proverbs 20:26 also says, **"A wise king scattereth the wicked,** and bringeth the wheel over them" You must ride in the chariot of your kingship to scatter and destroy the wicked. It is out of your kingship that you wage war, not in your identity as a priest. Not understanding this concept is what has made many spiritual warfare books ineffective in producing lasting victory in the lives of saints. Most of these books do not teach saints to wage war out of their identity as kings.

When you understand that you are a king and that it is out of your kingship that you wage war, you will begin to experience amazing victories over the enemy in your life. Waging war as a King is what makes this book effective. So, throughout this book, I have royal decrees for you to declare. As a king, you have a royal sentence in your mouth that cannot be countermanded by the kingdom of darkness. **"Where the word of a king is, there is power: and who may say unto him, What doest thou?"** (Eccl 8:4). Devils cannot resist the decrees, and sentences of a king.

To wage war, as a king, you need training. David, one of the greatest warrior kings said, **"Blessed be the LORD my strength, which teacheth my hands to war, and my fingers to fight"** (Psa 144:1). The purpose of this book is to teach your hands to war and your fingers to fight with 7 extremely powerful weapons of war in God's armory. With this training, you will destroy demonic world rulers, principalities, and powers, and set free entire cities and nations!

HEAVEN'S ARMORY OF WEAPONS

> *Hast thou entered into the treasures (storerooms) of the snow? or* ***hast thou seen the treasures (store-houses) of the hail, Which I have reserved against the time of trouble****, against the day of battle and war?* (Job 38:22-23).
>
> ***The LORD hath opened his armoury****, and hath brought forth the weapons of his indignation... (Jeremiah 50:25).*

God as a man of war has an armory in heaven where He keeps His weapons of war. And in this book, God is opening up His armory to *you!* Every kind of weapon you can imagine is kept in heaven's armory *"... against the day of battle and war."* An armory is where an army keeps a supply of weapons for times of war. It is also a weapons factory where weapons are invented, designed, and manufactured. King Solomon for example built cities to store his weapons and chariots. The Bible says Solomon "... **had a thousand and four hundred chariots, and twelve thousand horsemen, whom he bestowed in the cities for chariots,** and with the king at Jerusalem." (1 King 10:26) Every believer goes through days, times, and seasons of battle and war and for this purpose, God has reserved weapons customized for any kind of battle you face in His heavenly armory! You have access to these storehouses because you are a joint heir with Jesus of all that God has! That is why in Ephesians Paul says,

---

> *Finally, my brethren, be strong in the Lord, and in the power of his might.* ***Put on the whole armour of God,*** *that ye may be able to stand against the wiles of the devil* (Ephesians 6:10-11).

The word "armour" is the Greek word "hoplon" which according to the Strong Concordance means "weapons of warfare". "Hoplon" is the same word used for ***"weapons"*** in 2 Corinthians 10:4! So, God wants us to be equipped with ***all*** the instruments and weapons of war in His armory. He wants you to put them on, that is why He says, "Put on the whole armour (all the weapons) of God." Paul lists six of these weapons in Ephesians 6, and those are mainly defensive. ***There are many other weapons that the Holy Spirit reveals throughout the Word of God that you need.*** In this book, you are going to access seven of these weapons! God made you in His image and likeness so you can use His weapons! And you are not just to use His weapons in battle, YOU ARE GOD's Battle-ax and Weapons of War!

Thou art my battle axe and weapons of war... (Jeremiah 51:20).

That is what God says of you. You are God's battle-ax on earth! You constitute God's weapons of war on earth and in your region! That is why God wants to co-labor with you to win battles on earth by using you as His battle-ax and weapons of war. And with you, God will destroy demonic kings and princes and tear down the entire infrastructure of satanic fortifications in major regions all over the world. God is invading the cities and nations of the world with the Kingdom of Heaven and He is commandeering you into this battle engagement. That is why He girds and equips you for battle with His armor! For you are to be strong in *"the Lord, and in the power of HIS might"* (Eph 6:10).

In this book, I cover just seven of these divine weapons of mass destruction. At another time I will share on many other weapons in the bible. (Watch out for it!) So, this book is just to stir up the body of Christ to go hunting in the Word of God to find these weapons. Every weapon you find in the bible has divine power to destroy the kingdom of darkness. It is the glory of God to hide a thing and the glory of a king to seek them out (see Pro 25:2). Your glory as a King is to seek out these weapons and put them to work!

These are the Seven Divine Weapons of Mass Destruction you are about to learn:

1. The Morning Star
2. The Sword of the LORD
3. The Arrow of God
4. The Iron Rod
5. The Cup of the LORD's Fury
6. The Line and Plummet
7. The Cross

CHAPTER 1 - The Morning Star!

The Morning Star!

THE MORNING STAR IS ONE of the greatest weapons of mass destruction in the King's Armory. The Morning Star is the weaponization of the period in the Bible called the fourth watch which is from 3 am to 6 am or till the break of day (See Matthew 14:25; Mark 6:48). The Fourth Watch is a High Time in the spirit when you experience your greatest victories. The Morning Star is also known as the planet Venus that gives off that blue light you see before the Sun comes out in the morning and the twilight when the sun sets in the evening.

According to the European Southern Observatory, **"The orbit of Venus which lies inside the Earth's orbit takes it closer to Earth than any other planet. So Venus can be very easily seen from the Earth. It is often called the Morning or Evening Star and it is often the brightest object in the sky, with the exception of the Sun and Moon."** Do you see how close Venus is to the earth? That shows you the great influence the Morning Star has over what happens on the earth.

A great mystery of the Morning Star is that it is the ruling celestial or heavenly body that divides the day from the night. "... Let there be lights in the firmament of the heaven **to divide the day from the night ...**" (Gen 1:14). The Sun rules the day, the Moon rules the night, and the morning star is the heavenly body that divides them. The Morning Star is like an umpire or referee that governs the sun and moon and determines how the day springs and sets. The morning star divides the sun and moon with the dayspring at dawn and with the twilight in the evening. That is what makes the Morning Star so powerful.

And I will give him the morning star. (Revelation 2:28)

The Morning Star operates mainly from 3 am to 6 am to transition the night into day. It is a period of heightened spiritual activity because that is when God visits the earth. The Bible says, "**... his going forth is prepared as the morning**" (Hos 6:3) The word "prepared" in the Hebrew is "kûn" which means to be firmly established. From the foundation of the earth, God has firmly established a daily visitation to man on earth every morning as Job 7:18 also states, **"And that thou shouldest visit him every morning ..."** It is God's presence on the earth at this time that makes the Morning Star one of the greatest weapons you can ever have as a believer.

During His earthly ministry, Jesus met with His Father in prayer at this time. **"And in the morning, rising up a great while before day,** he (Jesus) went out, and departed into a solitary place, and there prayed." (Mark 1:35) Both the Kingdom of God and the Kingdom of Darkness make use of the Morning Star. Satan was originally known as Lucifer, son of the morning or son of "Shachar" which means dawn (Isa 14:12). And so, he knows about the powers of the Morning Star. But Jesus Christ ***The*** Bright and Morning Star is far greater than a "son" of the morning (Rev 22:16). We the saints of God therefore have the upper hand in the use of this powerful weapon.

EXERCISING DOMINION WITH THE MORNING STAR

The morning star is the ruling rod of dominion as the Bible says, **"... and the upright shall have dominion over them in the morning"** (Psa 49:14). The fourth watch is therefore the best time to walk in dominion over the powers of darkness. Isaiah 26:9 states "with my spirit within me will **I seek thee early: for when thy judgments are in the earth,** the inhabitants of the world will learn righteousness." The fourth watch is when God's Judgments are on the earth! That is why God commands us in Jeremiah 21:12 to **"... Execute judgment in the morning."** No matter what the enemy has engineered, you can rise up in the morning to execute judgment and tear it all down!

---

*And that, knowing the time, that now **it is high time** to awake out of sleep: for now is our salvation nearer than when we believed. The night is far spent, the day*

> *is at hand: **let us therefore cast off the works of darkness, and let us put on the armour of light*** (Romans 13:11-12).

I love the way the KJV translates the fourth watch as a "high time." The night being far spent and the day being at hand is a description of the fourth watch of the night, and dawn, the time before the sun rises. It is within that period that God empowers you to "cast off the works of darkness" that the devil programmed in the night against you! That is also the right time to put on the armor of light. Armor is the protective gear of soldiers or warriors, which means that the best time to get up and wage war in the spirit is in the High Time of the fourth watch.

THE SHADOW OF THE MORNING

> *The murderer rising with the light killeth the poor and needy, **and in the night is as a thief.** The eye also **of the adulterer waiteth for the twilight,** saying, No eye shall see me: and disguiseth his face. **In the dark they dig through houses, which they had marked for themselves in the daytime:** they know not the light. <u>For the morning is to them even as the shadow of death:</u> if one know them, they are in the terrors of the shadow of death* (Job 24:14-17).

Job chapter 24 is a great expose on the works of darkness. In that chapter, Job talks about how the murderer, the adulterer, and the thief operate in the night against the Saints. But in Job 24:17 we discover that the morning star is like the shadow of death to them. The word "morning" is the Hebrew "bôqer" which means the dawn or the break of day. The workers of darkness know that if you rise at the break of day to confront them, they are sure to lose the battle! You see, evil workers, labor throughout the night, and then in the morning, they execute what they schemed in the night.

> *Woe to them that devise iniquity, and work evil upon their beds! when the morning is light, they practise it,* ***because it is in the power of their hand*** (Micah 2:1).

Evil workers devise iniquity and work evil in the night upon their beds. The power of darkness in their hand is energized in the morning. "When the morning is light, they practice it ..." They also know about the power of the morning. You see them asleep in their bed but they are out and about astral projecting all over the world breaking into houses they marked out in the day and attacking people in their sleep. (see Job 24:13-16) Vigilant saints dream of these activities and wake up with foreknowledge to confront and destroy those works. Jesus said,

---

> ***But while men slept, his enemy came and sowed tares among the wheat,*** *and went his way* (Matthew 13:25).

If you do not do anything about what the enemy programmed against you then what they planted in your life during the night will germinate and spring up during the day to trouble you. But you can rise in the morning and dismantle all those demonic programs. Do that now with these decrees, **O Morning Star cast off the works of darkness from my life in Jesus' name! O Morning Star be a shadow of death to the wicked in Jesus' name!** When you confront the works of darkness with the morning star, you become a shadow of death to the enemy.

MORNING STAR POWERS

Jesus said in Revelation 2:28 **"And I will give him the morning star."** Jesus is saying to you right now reading this book ***"I am giving you the Morning Star!"*** Everything Jesus has power to do with the morning star you have power in His name to do! You must now search everywhere in the Word of God where the Morning Star is mentioned to understand how to walk in this power. There is no other place in the Bible that more comprehensively covers these powers than in Job 38:12-15.

> *"**Hast thou commanded the morning since thy days; and caused the dayspring to know his place**; That it might take hold of the ends of the earth, that the wicked might be shaken out of it? It is turned as clay to the seal; and they stand as a garment. And from the wicked their light is withholden, and the high arm shall be broken"* (Job 38:12-15).

I encourage you to meditate on Job 38:12-15 and start using each verse as a decree to subdue the powers of darkness. In this passage of scripture, God reveals certain Morning Star powers. I will now briefly explain these powers and share with you how to wage war with them.

COMMANDING THE MORNING STAR

> *Hast thou **commanded the morning** since thy days; and caused the dayspring to know his place* (Job 38:12).

When God asks a question, it is not to embarrass you, it is to reveal an area of ignorance in your life and an impartation He is giving you. That whole question in Job 38:12 was an activation of a dormant creative power of dominion in Job. When God said "Hast thou commanded the morning" He was telling Job *"Do you know you can command the morning?"* That means Job did not know about commanding the morning. I was so surprised when the Spirit first revealed this truth to me in 2010. I said to myself, *"So I can command the morning star?"* From that day I began to harness the powers of the Morning Star in my prayers and that is when my life began to turn around in a major way.

You have this power to command the morning star just as Joshua commanded the Sun and the Moon to stay in place and they did. The bible says, "And there was no day like that before it or after it, **that the LORD hearkened unto the voice of a man"** (Jos 10:14). God listened to the voice of Joshua and did what he desired. Isaiah 45:11 states, **"Thus saith the LORD ... concerning the work of my hands command ye me."** You have the power to command whatever God created and formed with His hands!

> ***I have made the earth, and created man upon it: I, even my hands, have stretched out the heavens, and all their host have I commanded*** (Isaiah 45:12).

With His hands, God created the earth, man, the heavens, and all the heavenly bodies. He also said, **"... all their host have I commanded."** God created everything with a command, and with a command, you can rule and exercise dominion over what He created. That includes the morning star! Jesus said, "Verily, verily, I say unto you, **Whatsoever ye shall ask the Father in my name, he will give it you**" (John 16:23). You can ask for the powers of the Morning Star and the Father will grant them to you.

You can have the morning star as Jesus said in Revelation 2:28 "And I will give him the morning star." When you command the morning star it will obey and execute your commands! When you command the morning, your dayspring, your "shachar" will get back in place! "**... and caused the dayspring (shachar) to know his place.**" If your dayspring is out of place, a lot of things in your life will be out of alignment with God's will. This can cause confusion and chaos in your life. But as you command the morning, things will begin to line up in your life and you will begin to prosper and succeed in every area of your life!

THE HANDS OF THE MORNING STAR

> *"**That it might take hold of the ends of the earth,** that the wicked might be shaken out of it?" (Job 38:13).*

The Morning Star Has Hands to shake the wicked! The Morning Star (bôqer) and Dayspring (shachar) have hands to ***"take hold"*** or to grab the ends of the earth and shake the wicked out of it. Hands represent power in the spirit. In the fourth watch, the morning star infuses your hands with power to shake the wicked out of your life wherever they are in the world. That gives you a great advantage in battle because the enemy is very crafty and can attack you from a remote region of the world. But with the morning star, you can confront and destroy these works of darkness because the Dayspring can reach anywhere in the world faster than you can imagine.

With the hands of the Morning Star you cast off not just the wicked but also their evil works as Romans 13:12 reveals, "The night is far spent, the day is at hand: **let us therefore cast off the works of darkness...**" Psalm 91:5-6 describes some of these works of darkness as the terrors of the night and the pestilence that walks in darkness. These works are engineered to oppress, terrorize, and afflict your life. But with the hands and power of the Morning Star you can grab hold of all these works of darkness and cast them off from your life just like you throw off a blanket when you wake up from sleep! That is how easily you render those works null and void in the Fourth Watch! Now use these Morning Star decrees to cast off these works. These decrees have power to destabilize demonic agendas and dismantle enchantments trying to cage and hinder you in life.

- **O Morning Star take hold of my life, my finances, my marriage, and my city, and shake out the wicked in Jesus' name!**
- **With the hands of the Morning Star, I shake off every assassination attempt on my life in Jesus' name!**
- **With the hands of the Morning Star, I shake off every bewitchment, enchantment, and divination afflicting my life in Jesus' name!**

THE SEAL OF THE MORNING STAR

> *It is turned as clay to the seal; and they stand as a garment.* (Job 38:14)

The Morning Star has the power of a seal. A seal identifies and establishes ownership and possession and preserves the structural integrity of that which is sealed. A seal preserves and protects whoever or whatever is branded with it. God's seal is a shield against demonic break-ins and ensures the authenticity and genuineness of a thing. In the Fourth Watch, as your life is being turned under the seal of the morning star whatever God is doing in your life is protected and preserved. The enemy cannot tamper with anything stamped with God's seal.

Here is an awesome excerpt from Albert Barnes' commentary on Job 38:14

> The passing of the light over the earth in the morning, **seems to be like rolling a cylinder-seal on soft clay. It leaves distinct impressions; raises up prominent figures; gives form and beauty to what seemed before a dark undistinguished mass (Albert Barnes Notes on the Bible, Job 38:14. Emphasis mine)**

This commentary explains exactly what happens early in the morning during the fourth watch. You see, the fourth watch is the most transformative, malleable, and flexible time where things can be easily turned around spiritually. As you harness the powers of the morning star in the fourth watch you seal your day with whatever you desire. Just like a seal rolling over a document or object, creating distinct impressions, you are programming your day for success and whatever was dead, incomplete, tarnished, destroyed, and indistinct in your life is quickened, resurrected, and restored to its original state! You are also preserving whatever you have obtained from God.

When the Jewish leaders went to Pilate and asked him to secure the tomb of Jesus, he granted their request by giving them a seal. *"So they went, and made the sepulchre sure,* ***sealing the stone****, and setting a watch"* (Matt 27:66). Pilate gave them a Roman seal to place on the tomb and a watch, which is a battalion of Roman Soldiers to guard the tomb. When you receive the seal of the morning star, God brands you with His imperial signet ring and releases legions of angels to protect, preserve, and secure your life. The morning star is God's signet ring! Every time God visits you in the morning, He stamps and seals your life with His will, plans, and purposes.

When a seal is stamped on or rolled over an object, a wax of one's signature, sign, or symbol of authority is placed on the object and the wax hardens to create a solid impression. Whatever therefore that happens during the day is sealed in the fourth watch! With the morning star, you can program certain words into your life that will give you victory over the enemy throughout the day! For example, you can say, **"O Morning Star render every enchantment against my life useless in Jesus' name!** When you seal yourself with this morning star command you can be certain that no enchantment can work against you that day.

Another decree you can make is **"O Seal of the Morning Star prosper the work of my hands and establish my goings today in Jesus' name!"** That simple command will secure and establish whatever you do that day. Whatever you touch will prosper. Nothing can obstruct God's plans for you that day. Even if the enemy planned a disaster for you that day you can be sure that the morning star seal has been rolled over the day obliterating every trace of the enemy's plans and replacing them with God's will and plans for you that day! Halleluyah!!

THE BLINDING POWER OF THE MORNING STAR

> *And from **the wicked their light is withholden, and the high arm shall be broken*** (Job 38:15).

The morning star has blinding power! When the Morning star arises in your life it blinds the enemy. It turns whatever light they had in your life into darkness. You see, the kingdom of darkness tracks our movements constantly as David said, **"The wicked watcheth the righteous,** and seeketh to slay him" (Psa 37:32). The enemy is watching you to destroy you. All saints are under this demonic high-level surveillance. Concerning these evil watchers, David also says,

> *... they hide themselves, **they mark my steps,** when they wait for my soul. (Psalm 56:6).*

> ***They have prepared a net for my steps** ... they have digged a pit before me* (Psalm 57:6).

Jeremiah also spoke about how the enemy monitored him, **"All my familiars watched for my halting,** saying, Peradventure he will be enticed, and we shall prevail against him, and we shall take our revenge on him" (Jer 20:10). Jeremiah calls those monitoring his life **"familiars"**. They were familiar with him and with his life but had evil intentions against him. All those marked with a special ministry go through this. Even Jesus our Lord went through this.

> ***And they watched him, and sent forth spies,*** *which should feign themselves just men, that they might take hold of his words, that so they might deliver him unto the power and authority of the governor* (Luke 20:20).

That is the activity of familiars! They monitor you to catch you at a weak moment and then attack you. But God has an answer for every trick and device of the enemy. The Morning star is God's grace and power in your life to blind the demonic "familiars" of your life. The bible says, **"And from the wicked their light is withholden..."** (Job 38:15). When you keep the fourth watch, any insight and light the "familiars" have about your life is put out. The morning star has blinding power to blow out the candle of the wicked ones as Job 21:17 states, "How oft is the candle of the wicked put out!" Make this decree right now while this revelation is hitting your spirit.

- **O Morning Star, blow out the candle of the wicked in Jesus' name!**

Blindness is a major war strategy and weapon in God's arsenal. When the men of Sodom surrounded Lot's house, the angels blinded them. And when the Syrian Army with their fierce chariots and horses surrounded Elisha, he struck them with blindness. The Bible says;

> *In that day, saith the LORD, I will smite every horse with astonishment, and his rider with madness ... **and will smite every horse of the people with blindness*** (Zechariah 12:4).

With the Morning Star, you can strike demonic horsemen, watchers, and workers of iniquity with blindness and confusion. Paul walked in this power to blind Elymas the Sorcerer. "And now, behold, the hand of the Lord is upon thee, **and thou shalt be blind, not seeing the sun for a season. And immediately there fell on him a mist and a darkness**; and he went about seeking some to lead him by the hand" (Acts 13:11). The very moment the sorcerer was blinded, the eyes of a governor opened, and he got saved. That is what happens when you walk in this power and strike the enemy with blindness!

Once you master all these revelations and mysteries of the Morning Star you will begin to experience unprecedented breakthroughs, success, and progress in your life. Your days will go better as you take control over each day with these morning star powers before it ever starts. Now start doing that by making these powerful morning star decrees.

- **O Seal of the Morning Star, preserve my life and destiny today in Jesus' name!**
- **By the power of the morning star, I strike every evil eye monitoring my life with blindness in Jesus' Name!**
- **O Morning Star strike all those marking my steps with confusion, madness, and blindness and make their way dark and slippery in Jesus' name!**
- **O Morning Star blind all sorcerers and diviners monitoring my life in Jesus' name!**
- **By the Power of the Morning Star, I exercise power over nations in Jesus' name!**

- **O Morning Star blind every evil strong man oppressing my nation and break their high arm in Jesus' name!**
- **O Morning Star take hold of the ends of my city and nation and shake out the wicked in Jesus mighty name!**

CHAPTER 2 – The Sword

THE SWORD OF THE LORD

The sword of the LORD,
and of Gideon (Judges 7:18).

THE SWORD IS THE MOST prolific weapon of war in the bible. After all, the Word of God is a double-edged sword. "For the word of God is quick, and powerful, and sharper than any twoedged sword" (Heb 4:12). In the Bible, the sword represents war (See Revelation 6:4; 19:11-16). Jesus said to His disciples, "Think not that I am come to send peace on earth: I came not to send peace, but a sword" (Matt 10:34). Jesus came down to the earth with a Sword to destroy the Kingdom of darkness and get back everything Satan stole from Adam.

Today, God is putting a sword in your hand to wage war against the enemy of your life and destiny! Jehovah has a sword, Jesus has a sword, the Holy Spirit has a sword and ***you*** have a sword! Now let me give you the example of Gideon who used the sword of the LORD to win a battle against three huge armies. Before the battle, Gideon said to his 300 men,

---

When I blow with a trumpet, I and all that are with me, then blow ye the trumpets also on every side of

> *all the camp, and say,* ***The sword of the LORD, and of Gideon*** (Judges 7:18).

Gideon commanded his men to issue a sword decree the moment he blows his trumpet. This sword decree "the sword of the LORD..." unsheathes the Sword and releases it against the enemy. And so, the moment Gideon's army shouted **"The sword of the LORD, and of Gideon"** confusion and pandemonium broke out among the three armies that had come against Israel and they began to fight each other! "... **the LORD set every man's sword against his fellow,** even throughout all the host" (Jdg 7:22). That is the Sword of the LORD! Three huge armies were destroyed by just 300 men when they issued the sword decree. You can put your name in this sword decree in place of the name, Gideon, and issue the decree to win great battles!

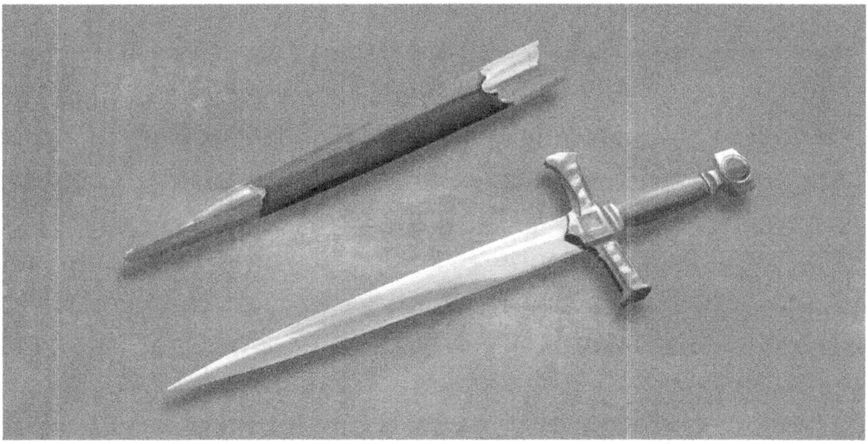

I the LORD have drawn forth my sword out of his sheath... (Ezekiel 21:5)

Now concerning the Sword of the LORD, the Bible says,

> *In that day* ***the LORD with his sore and great and strong sword*** *shall punish leviathan the piercing serpent, even leviathan that crooked serpent; and he shall slay the dragon that is in the sea* (Isaiah 27:1).

God's sword has three qualities – it is sore, great, and strong. It has power to slay the most powerful marine power of darkness, the dragon. Make this decree right now, **"O sword of the LORD, visit every demonic strongman and taskmaster oppressing my life in Jesus' name!"** That's right! The Sword of the LORD can be programmed to target a person, an entity, or a location just like a modern-day missile. Many highly advanced missiles have inbuilt computer systems that can be programmed to seek a target, evade intercepting missiles, and hit the desired target. Well, God had that technology before man ever did! Through this book, God is unsheathing His sword and activating it in your hand to empower you to fight His battles!

PROGRAMMING THE SWORD

> *"O thou sword of the LORD, how long will it be ere thou be quiet? put up thyself into thy scabbard, rest, and be still. How can it be quiet,* ***seeing the LORD hath given it a charge against Ashkelon,*** *and against the sea shore?* ***there hath he appointed it"*** (Jeremiah 47:6-7).

You can lay a charge on the sword of the LORD, which means you can program or appoint the sword to go to a specific location. In this scripture you see God laying a charge on His sword by programming it to destroy Ashkelon and its coastlands. The word "appointed" in the Hebrew is "yâ'ad" which means "to fix upon by appointment, to meet at a stated time, to direct". "Yâ'ad" also means " to appoint, assign, designate, meet at an appointed place". When you lay a charge on God's sword or when you give it an appointment, you are programming it with the coordinates of where the sword should go. You are targeting, directing, or designating the sword to go to a certain place! To win your city lay this charge on the sword of the LORD.

- **O Sword of the LORD, I give you a charge against the demonic strong man ruling my city. Go forth, locate, and destroy that strong man in Jesus' name!**

The Sword of the LORD is a weapon of mass destruction because it can destroy the infrastructure of the evil principality ruling an entire city, nation, or region. That is why God could send His sword to destroy Ashkelon and its coastlands! To destroy the wicked nation of Babylon, God issued this extremely powerful sword decree that you must master;

> ***A sword is upon the Chaldeans, saith the LORD,*** *and upon the inhabitants of Babylon, and upon her princes, and upon her wise men* (Jeremiah 50:35).

Any word that proceeds out of God's mouth is an eternal word that cannot be broken. The Most High God is the Highest King in the whole universe whose word cannot be countermanded. "**Where the word of a king is, there is power:** and who may say unto him, What doest thou?" (Eccl 8:4) And get this, the word of God in His mouth and in your mouth is the same! So, when you issue these commands, they have divine power to execute God's will. These commands will not return to you void. They will be accomplished and will prosper in hitting the designated targets (see Isaiah 55:11). You have power to call for the sword of the LORD to come upon a certain target as you see God doing in this scripture;

> *...**for I will call for a sword upon all the inhabitants of the earth,** saith the LORD of hosts* (Jeremiah 25:29).

You can also call for the sword of the LORD to come upon the strongman ruling your city by saying, "**A Sword is upon the demonic prince ruling my city!**" That is a decree! That is you saying what God is saying when He deploys His sword against His enemies. The angels of God assigned to your life are waiting for you to command them. "Bless the LORD, ye his angels, that excel in strength, **that do his commandments, hearkening unto the voice of his word**" (Psa 103:20). Angels listen and act on the word of God coming out of your mouth as if those words came out of God's own mouth because they hearken "unto the ***voice*** of his word". You put angels to work when you issue sword decrees!

SWORD FAITH COMMANDS AND ACTIONS

Now I will share with you how to wield the sword in the spirit using Ezekiel 21. In this passage, God shows Ezekiel how to deploy the sword of the LORD by acting out prophetically certain faith works or actions because **"... faith without works is dead"** (Jas 2:20). To perfect your faith you must add corresponding works or actions. When Abraham obeyed God by offering his son Isaac on the altar that was his faith working or his actions cooperating with his faith. I like the way the Good News Bible translates James 2:22, "Can't you see? His (Abraham's) faith and his actions worked together; his faith was made perfect through his actions." In your prayer warfare, your faith and actions work together for you to experience victory. I want you to see some of these "faith works" or "faith actions" that go with wielding the sword of the LORD.

Sword Faith Action 1:
SHARPEN AND POLISH THE SWORD

---

*Son of man, prophesy, and say, Thus saith the LORD; Say, **A sword, a sword is sharpened, and also furbished: It is sharpened to make a sore slaughter**; it is*

> *furbished (Polished) that it may glitter ... And he hath given it to be furbished, that it may be handled: this sword is sharpened, and it is furbished, to give it into the hand of the slayer* (Ezekiel 21:9-11).

According to this scripture, a sword is sharpened and furbished

1. To make a sore slaughter
2. That it may glitter
3. That it may be handled
4. To give it into the hand of the slayer

Before you go into battle, you must sharpen and polish your sword for it to be effective in your warfare. **A blunt sword needs to be sharpened and a rusty sword needs to be polished or furbished. Before you handle the sword, it must glitter and flash like lightning!** The word glitter means lightning, a bright and flashing sword. Your sword cannot be blunt and dull. It must be sharp and gleam brilliantly before you engage the enemy. Before and after every battle, warriors would sit down and sharpen their swords with a whetstone. And then they would furbish or polish the sword with a special kind of oil to keep the sword from getting rusty. A good warrior takes good care of his sword because,

---

> *If the iron be blunt,* ***and he do not whet the edge,*** *then must he put to more strength* (Ecclesiastes 10:10).

A blunt and dull sword makes you work harder and longer. God wants you to be "a new sharp threshing instrument having teeth" (Isa 41:15). A sharp and polished sword makes you more effective in battle. With just one swing of the sword, you decapitate the enemy and end the war!

WHET AND SHARPEN YOUR SWORD THROUGH PRAISE AND WORSHIP!

Let the high praises of God be in their mouth, and a twoedged sword in their hand (Psalm 149:6)

Praise and Worship should precede prayer warfare. High praises shift and reset the atmosphere and stir up the warfare anointing of a king on you. As high praises flow out of your mouth God goes up with a shout and your sword is sharpened and furbished in the spirit! David the great warrior King excelled at this. Before he engaged the enemy in battle he would praise and worship God and call on His name.

And he said, I will love thee, O LORD, my strength. The LORD is my rock, and my fortress, and my deliverer; ***my God, my strength,*** *in whom I will trust; my buckler, and the horn of my salvation, and my high tower.* ***I will call upon the LORD, who is worthy to be praised: so shall I be saved from mine enemies.*** (Psalm 18:1-3)

That was David sharpening and polishing his sword. David would take his harp and encourage himself in the Lord before ever stepping into battle. As he did this he would be empowered to do battle. "For thou hast girded me with strength unto the battle..." (Psa 18:39) This is the secret to David's invincibility in battle. He never lost a battle because he never fought with a blunt or rusty sword. He understood the wisdom of sharpening and polishing his sword to do a *"sore slaughter"*. As you employ this tactic in battle, I decree that you will always triumph in battle! No enemy will be able to stand against you when you confront them with a sharpened and furbished sword!

SHARPEN AND POLISH YOUR SWORD BY DWELLING IN THE PRESENCE OF GOD!

> ***he hath made my mouth like a sharp sword;*** *in the shadow of his hand hath he hid me,* ***and made me a polished shaft;*** *in his quiver hath he hid me* (Isaiah 49:2).

God makes your mouth like a sharp sword and you become a polished shaft as you abide, hidden away, in the presence of the Most High. Staying in the presence and face of the LORD sharpens you. As Proverbs 27:17 states, **"Iron sharpeneth iron; so a man sharpeneth the countenance of his friend."** This is why Moses' face shone when he came down Mount Sinai after spending 40 days and 40 nights with Jehovah. When you are sharpened and furbished in the presence of God you become more effective in battle. You consummate your battles with greater ease and speed.

SHARPEN YOUR SWORD BY STUDYING AND MEDITATING ON GOD'S WORD

> *For the word of God is quick, and powerful,* ***and sharper than any twoedged sword****, piercing even to the dividing asunder of soul and spirit, and of the joints and marrow....* (Hebrews 4:12).

Meditating on the word of God makes you sharper than ***any*** two-edged sword which can pierce through every obstacle and kill the enemy! The great mystery of this revelation is that you do not just handle the sword, you are that sword! You are God's weapon! The members of your body are weapons of war! According to Romans 6:13, you need to yield "... your members as instruments of righteousness unto God." The word "instruments" in the Greek is "hoplon" which means weapons of war. And so, every part of your body is a weapon! You see this with Jesus when the bible says that **"out of his mouth went a sharp twoedged sword: and his countenance was as the sun shineth in his strength"** (Rev 1:16). When the word of God dwells in you richly, in the spirit, your face will blaze brilliantly like the sun and sharp two-edged swords will issue out of your mouth to smite the nations (See Rev 19:15). And with the sword of the Lord, you will shake and turn a whole nation to God in one day (ref Isa 66:16).

Sword Faith Action 2:
CLAP YOUR HANDS AND STAMP YOUR FEET

> *Thus saith the Lord GOD;* ***Smite with thine hand, and stamp with thy foot, and say,*** *Alas for all the evil abominations of the house of Israel!* ***for they shall fall by the sword,*** *by the famine, and by the pestilence* (Ezekiel 6:11).

This scripture is a direct command of God. **"Smite with thine hand, and stamp with thy foot, AND SAY ..."** When you clap your hands and stamp your feet you are waging war! As you issue sword commands you perfect your faith by smiting or clapping your hands. As you do this your enemies will "fall by the sword". Make this decree right now.

- **As I clap my hands and stamp my feet, I command household witchcraft operating against my life to fall by the sword in Jesus' name!**

God wants you to clap your hands in warfare because power is hidden in your hand. As Habakkuk 3:4 states, "he had horns coming out of his hand: and **there was the hiding of his power.**" Horns represent royal power in scripture. As a king, there is royal power in your hands! And there are lightnings in your hands! The word "hand" in the Bible represents power and might. **"... in thine hand is power and might"** (1 Chr 29:12). And Job 36:32 states "By two palms He hath covered the light, And layeth a charge over it in meeting" (YLT). When the palms of your hands come together in a clap, electrical charges are unleashed! As you clap your hands in warfare, bolts of lightning, thunder, swords, and arrows shoot forth. Your enemy falls when you clap your hands because your hands are instruments of war. Your hands are being trained for war!

And why stamp your feet? There is power in your feet to possess every place you tread on as God said to Joshua, **"Every place that the sole of your foot shall tread upon, that have I given unto you,** as I said unto Moses" (Jos 1:3). Power to inherit by dispossessing or expelling the enemy is invested in your feet! It is with your feet that you tread on serpents and scorpions and all the power of the enemy as Jesus told his disciples in Luke 10:19. The bible also says, **"Thou shalt tread upon the lion and adder:** the young lion and the dragon **shalt thou trample under feet"** (Psalm 91:13). Your feet have great power to subdue the enemy because in the spirit they are encased in shoes of iron and brass (See Deut 33:25). And so when you clap your hands and stamp your feet, you are causing great confusion and destruction in the camp of the enemy.

DOUBLE THE CLAP TO INCREASE THE SWORD ACTION!

> *Thou therefore, son of man, prophesy, and smite thine hands together,* ***and let the sword be doubled the third time,*** *the sword of the slain: it is the sword*

of the great men that are slain, which entereth into their privy chambers (Ezekiel 21:14).

When you double the clapping of your hands you double the times the sword comes upon the enemy! God instructed Ezekiel to clap his hands again to double the intensity of the sword warfare. The Sword is doubled to issue the kill blow *"the sword of the slain."* It is also doubled to execute "great men." Greater ranks of the enemy are destroyed the more you clap. And the more you clap your hands, the sword "... entereth into their privy chambers." This means that the Sword of the LORD goes on a search and destroy mission to smoke the enemy out of their hideouts or privy chambers. "The strangers shall fade away, and be afraid out of their close (hiding or secret) places" (Psa 18:45). This is the power of the sword!

This happened in real life when Ezekiel obeyed God. Three times Nebuchadnezzar came against Judea in the days of three evil kings – Jehoiakim, Jeconiah, and Zedekiah. The last King, Zedekiah, tried to flee from Jerusalem with his family, but he was captured, his sons killed before his eyes and his eyes gouged out. This happened because the sword was doubled the third time in the hands of Ezekiel the prophet. God allows all this to play out in the bible to give you an example of what happens when you execute Sword commands and actions!

Sword Faith Action 3:
POINT THE SWORD AT THE ENEMY'S GATES

I have set the point of the sword against all their gates*, that their heart may faint, and their ruins be multiplied: ah! it is made bright, it is wrapped up for the slaughter* (Ezekiel 21:15).

When you point the sword at the gates of the enemy in the spirit, you will break through the gates and demolish demonic strongholds. The gate is where a battle is concentrated because once the gate is taken, the city is yours! "In that day shall the LORD of hosts be for ... strength to them that **turn the battle to the gate"** (Isa 28:5-6). When you point the sword of the LORD at the gates, the enemy will lose heart and the ruins of the devils' strongholds will be multiplied! Many regions are ruled by demonic strong men who hold captive masses of people in their strongholds. To rescue them you must first bind the strong man. Jesus said,

I have set the point of the sword against all their gates ... (Ezekiel 21:15)

> *No man can enter into a strong man's house, and spoil his goods,* ***except he will first bind the strong man****; and then he will spoil his house* (Mark 3:27).

The Gate is the point of entry into the strong man's house or palace which is also the stronghold. That is why the battle is so intense around the gate. But as you set the point of the sword against the gate of the city, the strong man will be broken and his stronghold demolished! That is what happens when you come upon the strong man and overcome him (Luke 11:21-23). You overcome the strong man with the sword! Then you can bind the strong man and set the captives free and divide the spoil! God gave you divine authority to take the gates of the enemy when He said to Abraham **"... thy seed shall possess the gate of his enemies"** (Gen 22:17). You are that seed of Abraham and so you have a divine mandate to possess the gates of your enemies. You possess those gates by employing the sword of the LORD.

LIST OF SWORD FAITH COMMANDS AND ACTIONS IN EZEKIEL 21

Here are nine commands and actions you can use in your spiritual warfare.

1. **Confronting the enemy** "Son of man, *set thy face toward Jerusalem,* and drop thy word toward the holy places, *and prophesy against the land of Israel"* (Eze 21:2).
2. **Drawing Out The Sword** "Thus saith the LORD; Behold, I am against thee, and will *draw forth my sword out of his sheath,* and will cut off from thee the righteous and the wicked" (Eze 21:3).
3. **Sending Forth The Sword** "... therefore shall *my sword go forth out of his sheath against* all flesh from the south to the north" (Eze 21:4).
4. **Sharpening and Polishing The Sword** "Thus saith the LORD; Say, *A sword, a sword is sharpened, and also furbished: It is sharpened to make a sore slaughter; it is furbished that it may glitter"* (Eze 21:9-10).
5. **Crying and Howling** *"Cry and howl, son of man: for* it shall be upon my people, it shall be upon all the princes of Israel" (Eze 21:12).

6. **Smiting The Thigh** "terrors by reason of the sword shall be upon my people: *smite therefore upon thy thigh"* (Eze 21:12).
7. **Prophesying and Clapping Your Hands** "Thou therefore, son of man, *prophesy, and smite thine hands together, and let the sword be doubled the third time, the sword of the slain:* it is the sword of the great men that are slain, which entereth into their privy chambers" (Eze 21:14).
8. **Pointing The Sword at the Enemy's Gates** *"I have set the point of the sword against all their gates,* that their heart may faint, and their ruins be multiplied..." (Eze 21:15).
9. **Thrusting the Sword to the Left and Right** *"Go thee one way or other, either on the right hand, or on the left,* whithersoever thy face is set" (Eze 21:16).

These are all prophetic acts and faith actions you can execute in your spiritual warfare. Now do not allow the sword to be quiet or to be still. How can the sword "... be quiet, seeing the LORD hath given it a charge against Ashkelon, and against the sea shore? there hath he appointed it" (Jer 47:7). Lay a charge on the sword and put it to work with these sword decrees;

- **I lay a charge on the Sword of the Lord against the enemy of my life in Jesus name**
- **I appoint the sword to destroy every evil strongman ruling my father's house and mother's house in Jesus' name!**
- **A sword upon every taskmaster oppressing my life in Jesus' name!**
- **A sword upon every witchcraft coven targeting my life in Jesus' name!**
- **A sword upon every stranger to my destiny in Jesus' name!**

- **As I clap my hands and stamp my feet let the enemy exacting upon my life fall by the sword in Jesus' name!**
- **As I clap my hands let the sword be doubled upon every dark power hindering my life in Jesus' name!**
- **As I thrust the sword to my right hand and to my left hand, I destroy every Sihon and Og occupying my land, in Jesus' name!**
- **By the power of the sword, I possess my possession in Jesus' name!**

CHAPTER 3 – The Arrow of God!

THE ARROW OF GOD

The arrow of the LORD'S deliverance, and the arrow of deliverance from Syria... (2 Kings 13:17)

THE BOW AND ARROW ARE ancient weapons used in wars to attack the enemy from a long distance. An outnumbered army can therefore decimate a larger army with just arrows. The Hebrew word for "arrow" is "chêts" which means a piercer, an arrow, a wound, and a thunderbolt. The Hebrew root of "chêts" is "châtsats" which means "to chop into, pierce or sever, to curtail, to cut off in the midst." A variation of that word is "châtsâh" which means to cut or split in two, to halve, to divide, and to reach unto. That is how powerful arrows are! Arrows have power to overtake, pierce, or wound a target and split it up! David in Psalm 45 says;

Thine arrows are ***sharp*** *in the heart of the king's enemies...* (Psalm 45:5).

Arrows were sharpened by archers to be able to pierce through thick armor and damage vital organs of the body. Let's see two examples of how arrows were used in the bible. For many years,

Ahab the wicked King of Israel, escaped the judgments of the Lord until an arrow caught up with him. "And **a certain man drew a bow at a venture, and smote the king of Israel between the joints of the harness ...**" (1 Kings 22:34). King Ahab had disguised himself as an ordinary soldier to deceive the Syrian army. But God had already decreed that he was to die in this battle. Ahab was a battle-tested warrior who won several battles but, in this war, an arrow released from the bow of an ordinary warrior stopped him. The arrow pierced through the *"joints of the harness"* cutting through Ahab's lungs and stomach. He died from this fatal wound.

Jehu was also a masterful archer. Having been anointed to destroy Ahab's household and dynasty, he set out in his chariot and attacked Jehoram who had succeeded his father Ahab. When Jehu confronted him with the wickedness and witchcrafts of Jezebel, Jehoram sensing danger fled in his chariot. But Jehu's arrow overtook him. "And Jehu drew a bow with his full strength, **and smote Jehoram between his arms, and the arrow went out at his heart,** and he sunk down in his chariot" (2 Kings 9:24). Jehu's arrow ate up the distance between him and Jehoram and killed him. That is the power of the arrow. It has power to reach a target from a long distance and pierce, divide, split and cut it up!

In the ancient world, archers would attach feathers to the shaft of the arrow to make it fly to its target. That is the glory of an arrow - It flies! It pursues, overtakes, and stops the enemy. David prophesied by the spirit and said, **"Thine arrows are sharp in the heart of the king's enemies;** whereby the people fall under thee" (Psalm 45:5). Because of how sharp an arrow is, it pierces and cuts through armor with devastating force causing the enemy to fall dead or to be mortally wounded.

Today, wooden arrows are replaced by missiles that can be launched over long distances to destroy certain targets. In December 2019, Russia deployed the Avangard Hypersonic Missile System. These hypersonic nuclear missiles travel twenty times faster than the speed of sound rendering interception virtually impossible! It is easier to understand spiritual truths when you first understand the natural (1 Cor 15:44). If hypersonic missiles can travel 20 times faster than the speed of sound how much more spiritual arrows?

GOD HAS ARROWS IN HIS ARMORY

> *And the LORD shall be seen over them,* ***and his arrow shall go forth as the lightning...*** (Zechariah 9:14).

Lightnings are God's arrows! Man uses arrows because God uses arrows. Psalm 64:7 says, **"But God shall shoot at them with an arrow**; suddenly shall they be wounded." God shoots arrows at His enemies. That means He has bows and arrows in His Armory! And God's arrows "... go forth as the lightning" (Zechariah 9:14). And you are the premier arrow in God's quiver!

> *"And he hath made my mouth like a sharp sword; in the shadow of his hand hath he hid me,* ***and made me a polished shaft (arrow); in his quiver hath he hid me"*** (Isa 49:2).

You are indeed a weapon in God's hand. That is why you see God bending the tribe of Judah like a bow and firing off the children of Ephraim like arrows against Greece (Zechariah 9:13). You are a projectile of God. You are the Arrow of God's deliverance for entire regions! He has made your mouth like a sharp sword! You are hidden and fashioned and polished and sharpened by God's own hand to fight His battles on earth. God's arrows are described as follows:

> *Sharp arrows of the mighty, with coals of juniper* (Psalm 120:4).

> ***For the arrows of the Almighty*** *are within me,* ***the poison whereof drinketh up my spirit: the terrors of God*** *do set themselves in array against me* (Job 6:4).

God's sharp arrows are tipped with poisonous coals of juniper which according to Matthew Henry's commentary on Psalm 120:4, "have a vehement heat, and keep fire very long." These arrows are kindled with the fires of hell to sap the strength and energy of the wicked! "... the poison whereof drinketh up my spirit" (Job 6:4). That is why Job calls these arrows **"the terrors of God."** God uses these arrows to execute His judgments on the wicked who are continually devising mischief against God's people.

> *God judgeth the righteous, and* ***God is angry with the wicked every day.*** *If he turn not, he will whet his sword;* ***he hath bent his bow, and made it ready.*** *He hath also prepared for him the instruments of death;* ***he ordaineth his arrows against the persecutors*** (Psalm 7:11-13).

In this scripture, you see that God patiently waits for the wicked to repent because He has no pleasure in the death of a sinner. In Revelation 2, you see Jesus giving Jezebel time to repent! But if the wicked "turn not" or repent, God bends his bows and strikes them with His arrows. God is angry with the wicked "every day" because He sees them destroying the lives of His people like one eats bread!

> *Have all the workers of iniquity no knowledge?* ***who eat up my people as they eat bread,*** *and call not upon the LORD* (Psalm 14:4).

God sees all that the workers of iniquity are doing! "The eyes of the LORD are in every place, beholding the evil and the good" (Pro 15:3). He sees them exchanging the lives of saints for wicked people (see Ezekiel 13:18-19). That is why God bends His bow, ordains His arrows, and suddenly shoots at them. Do you know what it means to ordain an arrow? It means God programs the coordinates of your enemy's location into the arrow. That is how God prepares arrows or His instruments of death against the enemy. I will soon teach you how He does this through prophetic targeting. As you read this book, I command the arrows of God to be ordained to scatter and destroy all the workers of insurrection plotting your demise, in Jesus name!

THE ARROWS OF THE WICKED

Satan is a powerful warrior versed in the art of warfare. In Revelations 12 you see him as a war general leading his angels to fight against Michael and his angels. Satan therefore knows and understands the use of arrows and he trains his earthly occultic ministers how to fire evil arrows of bitter words at people. These evil arrows cause infirmities, sicknesses, and even death.

> *Hide me from the secret counsel of the wicked; from the insurrection of the workers of iniquity: Who whet their tongue like a sword,* ***and bend their bows to shoot their arrows, even bitter words*** (Psalm 64:2-3).

Bitter words are spiritual arrows! And so the wicked have perfected the art of using their tongue like a bow to shoot arrows of bitter words, that is, curses, lies, slander, and accusations at the righteous. Jeremiah describes the wicked as those who **"bend their tongues like their bow for lies ... Their tongue is as an arrow shot out"** (Jer 9:3,8). This is what happens when workers of iniquity do their incantations and enchantments. They shoot evil arrows at the perfect and the upright with their words!

> *For, lo,* ***the wicked bend their bow,*** *they make ready their arrow upon the string,* ***that they may privily shoot at the upright in heart*** (Psalm 11:2).
>
> ***That they may shoot in secret at the perfect:*** *suddenly do they shoot at him, and fear not* (Psalm 64:4).

The perfect and the upright are the saints of the Most High. That includes you. But I decree that no arrow of enchantment and divination can prosper against you in Jesus' name! Make these decrees three times before you go any further:

- **I deflect every evil arrow fired at me with the Shield of Faith in Jesus' name**
- **Every arrow of infirmity in my body, be uprooted now, in Jesus' name!**

People of God are under constant attack from these arrows. For example, Psalm 91:5 talks about **"... the arrow that flieth by day"**. What protects you from these arrows is being in the presence of God, dwelling under the shadow of the Almighty, and staying in the word. "He shall cover thee with his feathers, and under his wings shalt thou trust: **his truth shall be thy shield and buckler"** (Psalm 91:4). May the wings of God shield you from evil arrows in Jesus' name! May His word be a shield and a buckler to you! Now, fire off this arrow decree seven times

- **Every evil arrow fired against my life, backfire now, in Jesus' name!**

The enemy also makes use of certain arrows called "fiery darts." This is just the enemy counterfeiting what God does with arrows of juniper. Fiery darts are arrows set on fire and launched against a target to burn it up. But God has given you the shield of faith to quench these flaming arrows. "Above all, taking the shield of faith, wherewith ye shall be able to quench all the fiery darts of the wicked" (Eph 6:16). Extinguish these fiery darts with this decree

- **I quench every fiery arrow fired against my life and destiny in Jesus' name!**

As you are holding this book in your hands, I command every arrow fired at you to be uprooted from your life and cut to pieces in Jesus' name! "**... when he bendeth his bow to shoot his arrows, let them be as cut in pieces**" (Psa 58:7). I command the bow and arrows in the hands of any evil archer targeting your life to be shattered in Jesus' name! "And I will smite thy bow out of thy left hand, and will cause thine arrows to fall out of thy right hand" (Ezek 39:3). Disarm evil archers with this juniper tipped arrow right now:

- **I smash the bow of the enemy and destroy his arrows in Jesus' name!**

THE ARROW OF THE LORD'S DELIVERANCE

Prophetic Targeting

Now Elisha was fallen sick of his sickness whereof he died. And Joash the king of Israel came down unto him, and wept over his face, and said, O my father, my father, the chariot of Israel, and the horsemen thereof. ***And Elisha said unto him, Take bow and arrows. And he took unto him bow and arrows.*** *And he said to the king of Israel, Put thine hand upon the bow. And he put his hand upon it: and* ***Elisha put his hands upon the king's hands.*** *And he said,*

> *Open the window eastward. And he opened it. Then Elisha said, Shoot. And he shot. And he said,* ***The arrow of the LORD'S deliverance, and the arrow of deliverance from Syria: for thou shalt smite the Syrians in Aphek, till thou have consumed them*** (2 Kings 13:14-17).

Elisha the prophet is sick and approaching death. During his prophetic ministry, he single-handedly delivered Israel from their enemy the Syrians. Several times God transported him in the spirit to listen in on the Syrian War Council. Armed with this divine intelligence, Elisha saved the king of Israel from many ambushments. One time, Elisha blinded an entire Syrian army sent to arrest him and then marched the blinded army to the King of Israel. From that day the Syrians never came again to Israel. And so, as Elisha lay dying on his bed, the King knew his nation was losing its defense system, the "*Chariot of Israel and the horsemen thereof*". That is how powerful Elisha the prophet was. This is the backdrop to the revelation of one of the most powerful divine weapons, the arrow of the Lord's deliverance.

As the desperate King cried, the prophet calmly instructed him to grab his bow and arrows and open the window towards Syria. Then laying his hand on the king's hand he commanded the King to shoot the arrow towards Syria. That is God's divine prophetic targeting system! In your prayer warfare, you need the coordinates of the enemy you are dealing with to be programmed into the arrow you are firing off. The Lord can reveal the identity of the enemy by speaking to you in a dream. If you pay attention to your dreams you will be armed with foreknowledge to counteract and overcome all the maneuvers and plots of the enemy.

The arrow of the LORD'S deliverance, and the arrow of deliverance from Syria (2 Kings 13:17)

DIVINE SYNERGY IN BATTLE

Elisha The Prophet and Joash The King

Now, both the King and Prophet fired off the arrow. That is why you must understand that you are both a king and a priest. You need both offices to deal effectively with the enemy. As a king, you judge and wage war. But as a priest, you go before God in prayer and receive divine authority against the enemy. Too many believers rush presumptuously into battle without first consulting the Lord. That is why there are many casualties in the body of Christ. The bible says,

> ***For by wise counsel thou shalt make thy war:*** and in multitude of counsellors there is safety (Proverbs 24:6).

As a priest, you enquire of the Lord in prayer and secure permission and authority over the enemy. It is in your office as a priest that you approach God and consult with him about your warfare. This is where you receive revelation about the enemy's strengths and operations, and this is where you secure divine help against the enemy. This is why Elisha the prophet had to put his hand on the king's hand to guide the arrow to the right target. The prophetic grace must guide your hand in battle. I am teaching your hands to battle! When Saul the king went into his last battle with the Philistines, there was no prophetic hand on his royal hand.

> ***And when Saul enquired of the LORD, the LORD answered him not, neither by dreams, nor by Urim, nor by prophets*** (1 Samuel 28:6).

Because God had rejected Saul, there was no prophetic hand on his hand when he went into battle against the Philistines. No dreams! No Urim! No Prophetic voice! That is why Saul lost that battle. He and many of his sons, including Jonathan, died in that battle. The King needs the prophetic hand on his hand to be effective in battle. Throughout the bible, you see Kings and prophets working closely together. Samuel the prophet anointed both Saul and David. David the king had Gad the prophet as his Seer. **"... the word of the LORD came unto the prophet Gad, David's seer"** (2 Sam 24:11). Elisha sent a prophet to anoint Jehu. In the New Testament, when John the Baptist as a prophet baptized Jesus, heaven opened, and the Holy Ghost came upon Jesus! Later John said of Jesus, *"Behold the Lamb of God!"* (John 1:36) That is prophetic targeting and prophetic endorsement.

For the arrow of the king to be effective against the enemy it must be prophetically programmed and targeted.

But no matter how powerful the prophet is, it is the king who goes out to battle. "... at the time when kings go forth to battle" (2 Sam 11:1). ***It Is The King Who Prevails In War!*** Yes, the prophet put his hand on the king's hand but it was Joash the King who drew the bow and shot the arrow towards Syria. This is the divine synergy that helped Israel win three wars against Syria.

In the book of Acts when two saints are sent out one of them is usually strong in the prophetic. And so you see Peter often paired with John a powerful prophet who wrote the book of Revelation. You also see Paul sent out with Silas, another mighty prophet (Acts 15:32). **God made you a King and a Priest so you can apply this divine synergy in your warfare.** Harnessing your priestly and prophetic office you engage God and determine the targeting coordinates for the arrow. Then as a King, you take the bow and arrow and with prophetic insight you fire off the arrows into the heart of the enemy! Now fire off this arrow decree seven times.

- ***I release the arrow of the Lord's deliverance, and the arrow of deliverance from every strongman assigned against my life, in Jesus name!***
- ***I loose the arrow of the Lord's deliverance against the strongman ruling and oppressing my city, in Jesus' name!***

PERSISTENCE IN PRAYER

When Elisha put his hand on the king's hand and asked him to shoot, the King obeyed and shot the arrow through the window. Then Elisha instructed him to fire his arrows into the ground.

> *And he said, Take the arrows. And he took them. And he said unto the king of Israel, Smite upon the ground.* ***And he smote thrice, and stayed.*** *And the man of God was wroth with him, and said,* ***Thou shouldest have smitten five or six times;*** *then hadst thou smitten Syria till thou hadst consumed it:* ***whereas now thou shalt smite Syria but thrice.*** (2 Kings 13:18-19)

The King shot only three arrows and stopped! How tragic!! There were still more arrows in the king's quiver that is why the prophet was so unhappy with the King. Warriors used quivers to keep and hold their arrows. An archer or warrior could have as many as 7 to 15 arrows in his quiver. If Joash had been diligent in shooting the arrows into the ground, he would have destroyed the Syrians! Now he could only win 3 battles because he shot just three arrows. And that is exactly what happened.

> ***Three times did Joash beat him,*** *and recovered the cities of Israel* (2 Kings 13:25).

Joash won three wars against the Syrians. Every arrow the King fired off represented a major victory over the Syrians. That is a weapon of mass destruction! Can you imagine what would have happened if he had fired off five or six arrows? This is why you must learn to persevere in prayer. The bible says, **"Pray without ceasing"** (1 Thess 5:17). The prayer of perseverance is what breaks you through into greatness. The early church in the book of Acts achieved greatness because they understood the power of persistent prayer. **"And they continued stedfastly ... in prayers"** (Act 2:42). The church today is weak because we have lost this art of prayer.

God promises to speedily avenge those who persist in prayer (See Luke 18:7-8). When Peter was in prison the church prayed ***without ceasing*** for him and God sent an angel to release him!

Elijah bowed in prayer seven times before the rain fell. Daniel persisted in prayer for 21 days before the archangel Gabriel broke through the barricades of the demonic prince of Persia. Jesus prayed the same prayer three times in the Garden of Gethsemane.

> *"And he left them, and went away again, **and prayed the third time, saying the same words**"* (Matthew 26:44).

You are never to stop praying until something happens! **Every arrow the King fired caused him to win a war. Every time you fire off an arrow you win a war! As you fire off the arrows of the LORD's deliverance, in the spiritual realm bolts of lightning and arrows will be launched into the very heart of the enemy!** Now let's do a prophetic faith act using Psalm 144:6, "Cast forth lightning, and scatter them: shoot out thine arrows, and destroy them." Clap your hands and stamp your feet as you fire off these arrows seven times

- **I scatter and destroy every demonic host and witch-craft coven terrorizing my life in Jesus' name!**
- **I scatter and destroy every evil confederacy conspiring against me in Jesus' name!**
- **I loose the arrow of the Lord's deliverance against every satanic taskmaster assigned against my life and ministry in Jesus' name!**
- **I loose the arrow of the LORD's deliverance against every demonic prince ruling my city in Jesus' name!**

CHAPTER 4 – The Iron Rod!

THE IRON ROD

And he shall rule them with a rod of iron (Revelation 2:27).

THE IRON ROD IS THE Scepter of a King. As a king unto God, you have a crown, you have a throne, you have garments of royalty and you have a rod in your hands to subdue nations and break their power. The scepter is called a Rod of Iron because iron is the sign and symbol of strength and warfare (See Dan 2:40). I use "rod" and "scepter" interchangeably because the iron rod is a scepter. In Ezekiel 19:11, God describes Israelite Kings and Princes as having "... **strong rods for the sceptres of them that bare rule...**" That word rule in the Hebrew is "mâshal" which means to rule, reign, have dominion, or to have power. Therefore, the Iron Rod is a weapon that gives you ruling power as Revelation 12:5 says, "And she brought forth a man child, **who was to rule all nations with a rod of iron...**" God gives you the scepter or the rod to execute His kingdom on earth. The Bible says,

---

> ***But unto the Son he saith,*** *Thy throne, O God, is for ever and ever: a sceptre of righteousness is the sceptre of thy kingdom* (Hebrews 1:8).

The iron rod is called a **"sceptre of righteousness"** because it brings things into right standing with God's kingdom. Anything out of alignment and in opposition to God's righteousness is subdued by the iron rod. You must walk in your identity as a king to use this weapon because the iron rod is the scepter of God's Kingdom. And as a Joint Heir with Christ, you have access to this rod because, **"... as he is, so are we in this world"** (1 John 4:17). We are the body of Christ on the earth. Jesus is the head and we are the body and so whatever He walks in we walk in. Whatever He is, we are. And whatever He has, we have! This is why Jesus says

---

> *And he that overcometh, and keepeth my works unto the end, to him will I give power over the nations:* ***And he shall rule them with a rod of iron;*** *as the vessels of a potter shall they be broken to shivers:* ***even as I received of my Father*** (Revelations 2:26-27).

The Rod of Iron is a weapon of mass destruction because it gives you power over nations to rule them. Jesus Himself has this rod. (See Psa 2:9, 110:2; Rev 19:17) And Jesus wants you to have what the Father gave Him that is why He said, **"... even as I received of my Father."** Just as Jesus received the rod of iron from the Father, He is giving you the iron rod to rule the nations. In Revelation 1:6 Jesus makes you a king and in Revelation 2:27 He gives you a scepter to rule and subdue nations. ***You are a king with a scepter your hands!***

SWALLOWING THE ROD OF THE WICKED

> ***For the rod of the wicked*** *shall not rest upon the lot of the righteous* (Psalm 125:3).
> *Violence is risen up into* ***a rod of wickedness...*** (Ezekiel 7:11).

The enemy also has wicked rods to oppress the people of God. When Moses went into Egypt with the Rod of God, he had to confront two sorcerers, Jannes and Jambres who were the counselors of Pharaoh (see 2 Timothy 3:8). Jannes and Jambres used their rods to oppress the children of Israel and manipulate them into a bitter life of slavery and hard labor. Pharaoh ruled and oppressed Israel with the counsel of these sorcerers until Moses arrived in Egypt with the Rod of God in his hand to confront them.

With the Rod of God in his hands, Moses brought a new manifestation of God's kingdom on the earth. And so, not only Moses's rod but also Aaron's rod was infused with power to do incredible miracles. The Rod of Moses did miracles! Aaron's rod did miracles! And by this revelation, may the rod in your hand do miracles! And so, in many of the miracles Moses did, God simply instructed Moses to tell Aaron to use his rod. That is why when Moses and Aaron stood before Pharaoh, Moses told Aaron to throw his rod down. And guess what happened, those sorcerers also threw down their rods. But Aaron's rod swallowed their rods!

> *For they cast down every man his rod, and they became serpents:* ***but Aaron's rod swallowed up their rods*** (Exodus 7:12).

The demonic influence Pharaoh had over Israel was broken when Aaron's rod swallowed up the wicked rods! In the same way, may your rod swallow up the rod of the wicked! You see, Satan has given his servants' rods of wickedness to steal kill, and destroy. But according to Psalm 125:3 **"... the rod of the wicked shall not rest upon the lot of the righteous"** And so you can boldly decree and say,

- **The Rod of the Wicked will not rest on my inheritance in Jesus' name!**
- **O Rod of God, swallow up every wicked rod working against me, in Jesus' name!**

These Iron Rod decrees will stop the enemy from walking off with your inheritance.

SMASHING EVIL POTS WITH THE IRON ROD

> ***Thou shalt break them with a rod of iron;*** *thou shalt dash them in pieces like a potter's vessel* (Psalm 2:9). *Thus saith the LORD of hosts;* ***Even so will I break this people and this city, as one breaketh a potter's vessel, that cannot be made whole again...*** (Jeremiah 19:11).

The Iron Rod smashes nations as a potter smashes pots. When a pot is smashed with the Iron Rod it cannot be reconstituted, it "cannot be made whole again". Pots and cauldrons are used in the demonic realm to oppress people (see Micah 3:3). When God revealed to Jeremiah the Babylonian Invasion, He showed him a boiling pot from the north (see Jer 1:13-15). The boiling pot represented the Babylonian siege that destroyed Israel. Evil workers in Israel also called the city of Jerusalem a pot (see Eze 11:3-7).

Thou shalt break them with a rod of iron; thou shalt dash them in pieces like a potter's vessel (Psalm 2:9).

You see, a whole region, city, or nation can become a pot in the spirit. The strongman ruling a city in the spirit can oppress the people in that land with a pot of affliction. But with the Iron Rod, you can break evil strongmen and dash their strongholds "... in pieces like a potters vessel". That is why God is raising up kings with rods of iron in their hand to smash these pots into pieces and release the prisoners from the house of the strongman! Moses did this when he went to Egypt. For centuries Egypt was an iron furnace and firepot oppressing Israel until Moses came with a rod to confront Pharaoh. The Rod is the weapon of choice for dealing with demonic pots oppressing God's people. Now make this decree,

- ***I smash every evil pot oppressing my life with the Iron Rod in Jesus' name!***

And Moses stretched forth his rod toward heaven: and the LORD sent thunder and hail, and the fire ran along upon the ground; and the LORD rained hail upon the land of Egypt. (Exodus 9:23)

THE ROD OF GOD IN MOSES' HAND

> *... and **Moses took the rod of God in his hand.** And the LORD said unto Moses, When thou goest to return into Egypt, see that **thou do all those wonders before Pharaoh, which I have put in thine hand ...*** (Exodus 4:20-21).

The Rod of God in Moses's hand represented the mighty hand and outstretched arm of God. God told Moses exactly what the Rod in His hands will do. "**And thou shalt take this rod in thine hand, wherewith thou shalt do signs**" (Exodus 4:17).

The Rod is a Scepter of Signs and Wonders that executes God's judgment on Satan's kingdom! Moses's rod was infused with sovereign power. It was a conduit of divine wrath. Before Moses met God, it was just a shepherd rod. But that stick in his hand became the rod of God after the burning bush experience! With this rod, Moses struck and subdued Egypt with several plagues. Now let us see how Moses used the rod to subdue Egypt

Aaron throws down his rod before Pharaoh. "Take thy rod, and cast it before Pharaoh, and it shall become a serpent" (Exo 7:9). When Aaron's rod became a serpent, the sorcerers of Pharaoh also threw down their rods and the rods became serpents. But Aaron's rod swallowed their rods just as Jesus swallowed up death in victory! Now the sorcerers had no scepter to exercise dominion over the Israelites. Their power was broken! You break the ruling power of the enemy when you throw down your rod and swallow up their rods!

Moses strikes the waters of Egypt with the rod. God commanded Moses to confront Pharaoh at the River Nile, "Thus saith the LORD, In this thou shalt know that I am the LORD: behold, **I will smite with the rod that is in mine hand upon the waters which are in the river, and they shall be turned to blood.** And the fish that is in the river shall die, and the river shall stink ..." (Exo 7:17-18). God then instructed Moses to tell Aaron to stretch his rod over all the water bodies of Egypt. So, while Moses struck the main river body with the rod, Aaron stretched his rod over all the waters of Egypt and the waters turned to blood. Divine Synergy! **May the Iron Rod plague any marine power oppressing your life in Jesus' name!**

Moses and Aaron plague Egypt with frogs. "And the LORD spake unto Moses, Say unto Aaron, **Stretch forth thine hand with thy rod over the streams,** over the rivers, and over the ponds, and cause frogs to come up upon the land of Egypt" (Exo 8:5). Moses and Aaron did what God told them and frogs came out of the water to plague Egypt. You have this power in your hands! When you stretch your rod, creation will work with you to execute God's will.

Moses and Aaron plague Egypt with lice. "Stretch out thy rod, and smite the dust of the land, that it may become lice throughout all the land of Egypt" (Exo 8:16). When they stretched their rods and struck the land, dust was weaponized to plague Egypt with lice! You can work with God in the same way. You can command the elements and creation and they will obey you because creation fulfills the word of God. "Fire, and hail; snow, and vapour; stormy wind **fulfilling his word"** (Psa 148:8). You can do this because God dwells and walks in You (2 Co 6:16). And with God dwelling in you, there is nothing you cannot do.

Moses plagues Egypt with thunder and hail. "And **Moses stretched forth his rod toward heaven: and the LORD sent thunder and hail,** and the fire ran along upon the ground; and the LORD rained hail upon the land of Egypt." (Exo 9:23). Nothing like this had ever happened in Egypt. When Moses stretched the rod upwards, heaven responded! And now when ***you*** stretch your rod towards the sky, heaven will back your words.

The Iron Rod has power to deploy winds. "And Moses stretched forth his rod over the land of Egypt, and the LORD brought an east wind upon the land all that day, and all that night; and when it was morning, **the east wind brought the locusts"** (Exo 10:13). The east wind brought a huge swarm of locusts that destroyed the vegetation of Egypt. The Iron rod in your hand has power to dry up the financial resources of the enemy.

Moses divides the sea with the rod in his hand to make a way for the Israelites to pass over. God said to Moses, "But lift thou up thy rod, and stretch out thine hand over the sea, and divide it: and the children of Israel shall go on dry ground through the midst of the sea" (Exo 14:16). With the same rod, Moses brought the sea back on the Egyptians drowning the entire Egyptian army. You will also breakthrough when you stretch the Iron Rod over every impossible situation impeding your life.

The Lord wants you to walk in the power of the Iron Rod! He allowed Moses to walk in this power so you can understand how to manifest the amazing power of the rod. With this scepter in your hand, you have power to subdue nations with miracles of judgment like Moses did to Egypt. Right now the Lord is activating the Rod of Iron in your hand! You can do what Moses did with his rod! You can stretch the rod to plague and subdue the enemy, you strike with it, and stretch it towards heaven to commandeer the heavenly bodies. Creation will obey you as you walk in the power of the Iron Rod. Now, confront and destroy the rod of the wicked operating in your region with these decrees!

- **The LORD has broken the staff of the wicked and the scepter of rulers oppressing this city in Jesus' name!** *(Say this three times with authority)*
- **O Scepter of righteousness, arise and smash every evil principality and destroy all demonic strongmen ruling this region in Jesus' name!**
- **O Fire of God consume the rod of every evil prince operating in this region until they have no scepter to rule in Jesus' name!**
- **O Rod of God swallow up all wicked rods in this land in Jesus' name!**
- **With the Iron rod, I smash all strongholds of rebellion in this region and dash them in pieces like a potter's vessel in Jesus' name!**
- **With the Rod of God, I rule in the midst of mine enemies in Jesus' name!**
- **I decree that the rod of the wicked shall not rest on this city in Jesus' name!**
- **By the Power of the Iron Rod, I smash into pieces the rod of the strongman oppressing this region in Jesus name**
- **And with the Rod of Iron, I rule this region and loose the Kingdom of God into this land in Jesus' name!**
- **For thy Throne O God, is forever and ever: A scepter of righteousness is the scepter of thy kingdom!**

CHAPTER 5 – The Cup!

THE CUP OF THE LORD'S FURY

***Take the wine cup of this fury at my hand...** (Jeremiah 25:15)*

NOW, I WANT TO SHARE with you an extremely powerful weapon of mass destruction in God's arsenal, the Cup of the LORD's Fury! Sadly, extraordinarily little is known about this weapon in the body of Christ. Satan is terrified of this weapon and will do anything to prevent the church from knowing about this explosive weapon. This is because the cup of the Lord's fury is not just a weapon, it is a cocktail of weapons! The bible says, "Upon the wicked he shall rain snares, fire and brimstone, and an horrible tempest: **this shall be the portion of their cup"** (Psa 11:6). The Cup of the LORD's fury comprises

- Snares – these are traps and minefields in the spirit set to entrap the enemy
- Fire
- Brimstone (Burning Sulphur)
- A Horrible Tempest

What a weapon! The Cup of the LORD's fury is a massive bomb that rains snares, fire, brimstone, and a horrible tempest on the head of the wicked! The Hebrew word for Horrible is "zal'âphâh / zil'âphâh" which means the "raging heat of burning wind". When you release this weapon it causes a powerful explosion in the camp of the enemy. The color of the mixture in the cup is red that is why it is called the "wine cup".

> ***For in the hand of the LORD there is a cup, and the wine is red;*** *it is full of mixture; and he poureth out of the same: but the dregs thereof,* ***all the wicked of the earth shall wring them out, and drink them*** (Psalm 75:8).

The wine-cup is **"full of mixture"**. It is a combination of several weapons and the intended target of this weapon is "all the wicked of the earth." Can you imagine a weapon that can target *all* the wicked of the earth? That is a weapon of mass destruction! When you dispense this wine cup of the LORD's fury against the demonic regional powers of your city or nation, you will see the stronghold of darkness destroyed and all the souls bound under those demonic powers freed. This weapon is part of God's end-time judgments in the book of Revelation against all those that follow the antichrist and also against Mystery Babylon (See Revelation 14:10; 16:19; 18:6).

GOD TEACHES JEREMIAH HOW TO USE THE WINE CUP

The following passages describe how to use the Cup of the Lord's fury.

> *For thus saith the LORD God of Israel unto me;* ***Take the wine cup of this fury at my hand, and cause all the nations, to whom I send thee, to drink it. And they shall drink, and be moved, and be mad,*** *because of the sword that I will send among them.* ***Then took I the cup at the LORD'S hand, and made all the nations to drink,*** *unto whom the LORD had sent me: To wit, Jerusalem, and the cities of Judah, and the kings thereof, and the princes thereof, to make them a desolation, an astonishment, an hissing, and a curse; as it is this day* (Jeremiah 25:15-18).
>
> *And all the kings of the north, far and near, one with another,* ***and all the kingdoms of the world, which are upon the face of the earth:*** *and the king of Sheshach shall drink after them. Therefore thou shalt say unto them, Thus saith the LORD of hosts, the God of Israel;* ***Drink ye, and be drunken, and spue, and fall, and rise no more, because of the sword which I will send among you.*** *And it shall be, if they refuse to take the cup at thine hand to drink, then shalt thou say unto them,* ***Thus saith the LORD of hosts; Ye shall certainly drink*** (Jeremiah 25:26-28).

Read these passages carefully because a great mystery of global warfare is embedded in them. You must first understand that God was angry with Jerusalem and many nations of the world and He sent several prophets to tell them to repent. But they refused to repent and mistreated God's servants. Therefore

God in His wrath made Nebuchadnezzar, the King of Babylon, His golden cup to destroy these rebellious nations.

> ***Babylon hath been a golden cup in the LORD'S hand,*** *that made all the earth drunken: the nations have drunken of her wine; therefore the nations are mad* (Jeremiah 51:7).

Babylon was to be the wine cup of God's fury to destroy the nations. But to deploy this weapon of mass destruction, God needed Jeremiah. This is because when God created the earth, He put it in man's hands. "The heaven, even the heavens, are the LORD'S: **but the earth hath he given to the children of men."** (Psa 115:16). That is a spiritual law that God Himself will not break. God therefore needs ***you*** to work with Him to bring regional transformation to your city with these weapons I am teaching you. And so God told Jeremiah to take the cup and cause the nations to drink of it (see Jeremiah 25:15). I want you to see this with the eye of your spirit. As God stands before Jeremiah with a cup in His hand, He tells Jeremiah to take the cup from His hand and give it to the nations to drink. If God could do that with Jeremiah, He can do that with you! God could do this with Jeremiah because He gave him power over nations and kingdoms,

> ***See, I have this day set thee over the nations and over the kingdoms,*** *to root out, and to pull down, and to destroy, and to throw down, to build, and to plant* (Jeremiah 1:10).

God set Jeremiah over nations and kingdoms to uproot and destroy them. God will never command you to confront or attack something you do not have authority over. This is why God gives you power in your kingship to destroy the works of the devil. The wine-cup is one of these powers you have as a king to execute the Lord's judgments on the earth.

"Then took I the cup at the LORD'S hand ..." (Jeremiah 25:17)

Jeremiah therefore took the cup from God's hand, "**Then took I the cup at the LORD'S hand, and made all the nations to drink,** unto whom the LORD had sent me" (Jer 25:17). In the spiritual realm, all the nations were represented by the demonic princes that ruled them. In 1 Corinthians 2:9, Paul calls these demons **"princes of the world".** These princes rule entire regions and nations such as the Prince of Persia and the Prince of Greece (See Dan 10:13,20-21).

After Jeremiah took the cup, God led all these princes of the world to him and caused them to line up before him. God then told Jeremiah what would happen when the princes drink from the cup.

And they shall drink, and be moved, and be mad, *because of the sword that I will send among them* (Jeremiah 25:16).

All the princes and kings of the earth began to stagger and walk around confused and dazed like drunken men! A sword, a war, embedded in the wine cup was unleashed against them when they drank from the cup! All this happened before Nebuchadnezzar brought war to their borders. Everything that happens in the natural is first established in the spiritual realm! As Jeremiah dispensed the cup to them, God told him to say to the demonic princes:

> ***Drink ye, and be drunken, and spue, and fall, and rise no more, because of the sword which I will send among you*** (Jeremiah 25:27).

I can assure you that God will use you in this manner. God will raise you up as a divine king to render His judgments in this manner in your region. God is giving you the wine cup to give to the demonic principalities causing havoc in your region! And I am sharing with you exactly how you can use this weapon. By the Spirit of the Living God of Hosts, I am teaching your hands to battle and your fingers to fight! You are to say to the rulers of the darkness in your region, **"Drink ye, and be drunken, and spue, and fall, and rise no more, because of the sword which I will send among you!"** They are to drink of the cup of the Lord's fury and vomit out the people they are oppressing. They are to fall and never rise again!

OVERCOMING STUBBORN SPIRITS

> *"Thus saith the LORD of hosts; **Ye shall certainly drink**"* (Jeremiah 25:28).

You see, not all evil spirits are the same. Certain evil spirits will resist your commands because they are not the demons you normally deal with. Some demons are greater than the others because of the rank they have attained in Satan's kingdom. When Jesus commissioned the disciples to go cast out devils and heal the sick, they came across a dumb and deaf spirit that they could not cast out. Jesus cast out that devil and healed the boy who was being oppressed by that evil spirit. The disciples came to Jesus later that day and asked Him why they could not cast out the devil and Jesus said, **"This kind can come forth by nothing, but by prayer and fasting"** (Mark 9:29). Jesus said ***"This kind"*** because this was not a normal demon! The disciples had to pay the price of prayer and fasting to get more authority and power to overcome this spirit. In the same way, whereas all the demonic princes drank the wine cup at Jeremiah's command, others were a *"this kind"* who refused to drink. God told Jeremiah exactly what to do to force them to drink.

> *And it shall be,* ***if they refuse to take the cup at thine hand to drink****, then shalt thou say unto them, Thus saith the LORD of hosts;* ***Ye shall certainly drink*** (Jeremiah 25:28).

Ye shall certainly drink! That is the command of a king! When you issue commands like this the enemy can no longer resist you because **"Thou shalt also decree a thing, and it shall be established unto thee"** (Job 22:28). That is the power of your words as a king. You shall have whatsoever you say if you have faith in your words. ***Ye shall certainly drink!*** When Jeremiah issued this decree, those stubborn princes were forced to drink of the wine cup. This command *"Ye shall certainly drink!"* broke their resistance!

After Jeremiah gave the cup to all the nations of the world they could no longer resist Babylon. They became **"a desolation, an astonishment, an hissing, and a curse"** (Jer 25:18). Nebuchadnezzar conquered them not because of his military might, but because a prophet had forced the ruling demonic princes of all those nations to drink from the wine cup of the Lord's fury. Before your next evangelistic outreach, dispense the wine cup to the strongman ruling that region and you will see that whole village, city, or nation saved and turned to the Lord! Now fire off these wine cup arrows into the very heart of the strongman, principalities, powers, and territorial forces ruling your region!

- **In Jesus' name, I deploy the wine cup of the Lord's fury against the strongman, principalities, and powers of darkness ruling my city!**
- **By the power of the wine cup, I loose the rain of snares, fire, brimstone, and a horrible tempest against the evil strongman ruling my city in Jesus' name!**
- **O Strongman, oppressing this region, drink of the wine cup of the LORD's fury and be drunken in Jesus' name! Be moved, be mad, and vomit out the spoil of this city! Fall and rise no more because of the sword I appoint against you, in Jesus' name!**
- **To all you stubborn dark princes and powers in this city, I say to you, Ye shall certainly drink! Ye shall certainly drink!! Ye shall certainly drink of the wine cup of the fierceness of God's wrath in Jesus' name!**
- **And by the power of the wine cup, I command the strongholds of the enemy in this city to become a desolation, an astonishment, a hissing, and a curse in Jesus' name!**

CHAPTER 6 – The Plumbline!

THE LINE AND PLUMMET

The Lord stood upon a wall ... with a plumbline in his hand (Amos 7:7).

THE PLUMBLINE IS A WEAPON of mass destruction so powerful God only uses it for very severe judgments. He uses it to destroy and wipe something or someone out of existence. The Plumbline is a very unique tool in an architect or carpenter's toolbox. According to Encyclopedia Britannica, **"A plumb line is a light line with a weight (plumb bob) at one end that, when suspended next to a workpiece, defines a vertical line."**

And so, the terms Plumbline or Plumb bob can be used interchangeably to describe an instrument that determines whether constructions or landscapes are vertical or straight. This is to ensure that constructions adhere to the dimensions and specifications drawn up by architects, builders, surveyors, carpenters, masons, and all those involved in any type of construction. Without this instrument buildings and constructions will be crooked, bent over, too high, too low, or uneven. This is important to understand because the glorious church cannot come into full manifestation unless the way of the Lord is prepared or plumblined! That is why before every major move of God, the plumbline is used to prepare the way.

> *The voice of him that crieth in the wilderness,* ***Prepare ye the way of the LORD, make straight in the desert a highway for our God.*** *Every valley shall be exalted, and every mountain and hill shall be made low:* ***and the crooked shall be made straight, and the rough places plain: And the glory of the LORD shall be revealed,*** *and all flesh shall see it together: for the mouth of the LORD hath spoken it* (Isaiah 40:3-5).

John the Baptist was that voice crying out in the wilderness (see John 1:23). John was the plumbline in God's hand to "plumb" and prepare Israel to receive the ministry of Jesus. Therefore, all those who received John's ministry of repentance received Jesus (ref Luke 7:29-30). And so in Isaiah 40:3-5 you see that the "way of the LORD" must be prepared. The highways of the LORD must be made straight by plumb lining the landscape. As you apply the plumb line to the way of the LORD, valleys are exalted; mountains and high places are brought low; crooked places are made straight and rough places are leveled. It is only then that the "glory of the LORD shall be revealed" for all flesh to see what God is doing.

When you unleash the power of the plumbline, all mountains of imposition and impossibilities in your life will be removed! All demonic strongmen and powers who have exalted themselves against your destiny will be pulled down! All valleys of depression, low self-esteem, lack, and insufficiency will be raised up! All the crooked places of iniquitous patterns and oppressive cycles in your life will be straightened out! And all rough places, stumbling blocks, invisible traps, and snares in your path that hinder and obstruct your progress and trip you up in life will be removed! May the plumb line of God accomplish all this in your life in Jesus' name! The Plumb line also obliterates the host of darkness and demonic forces assigned to prevent you from coming into your destiny. Before you go any further make these decrees.

- **I stretch out the line of confusion over all household enemies and I release stones of emptiness on their camp in Jesus' name!**
- **I plumbline every obstruction of familiar spirits out of my life, in Jesus' Name!**

Behold, the Lord stood upon a wall ... with a plumbline in his hand (Amos 7:7).

THE PLUMBLINE IN GOD'S HAND

In Amos 7:7 an amazing scene unfolds, "... and, behold, **the Lord stood upon a wall made by a plumbline, with a plumbline in his hand."** God appears before the prophet holding a plumbline in His hand. That proves that the plumbline is a divine instrument of mass destruction that God Himself uses. God had shown Amos a series of judgments coming against Israel. He showed Amos a swarm of locusts destroying the land and then a judgment of fire devouring the land. Amos interceded for Israel and got God to suspend those judgments. Then God appears to him standing on a wall with a plumbline in His hand. This wall represents Israel and at the beginning of her existence, God used the plumbline to plant and build her into a mighty nation. However, centuries later Israel was no longer aligned with what God plumblined her to be.

> *And the LORD said unto me, Amos, what seest thou?* ***And I said, A plumbline.*** *Then said the Lord,* ***Behold, I will set a plumbline in the midst of my people Israel:*** *I will not again pass by them any more:* ***And the high places of Isaac shall be desolate,*** *and the sanctuaries of Israel shall be laid waste; and I will rise against the house of Jeroboam with the sword* (Amos 7:8-9).

The same plumbline God used to establish Israel as a nation was going to be used to destroy her! This is because Israel had forsaken God. The children of Israel were at this point living in complete apostasy, serving Baals, and committing great sins against God. God had patiently endured their sins, but now he had come with the plumbline in His hand to judge Israel for her evil ways. He tells Amos, "Behold, I will set a plumbline in the midst of my people Israel: **I will not again pass by them any**

more." The Plumbline represents a final judgment that lays waste to a nation, a people, and a land.

When God pulls the plumbline out of His armory and sets it *in the midst of* a land it means utter destruction and desolation! There is a dreadful finality to this instrument of death. You must also set the plumbline in your region to destroy and lay waste to the stronghold of darkness and to demolish those dark places and haunts of cruelty. When you do that, all of heaven will begin to fight and contend with those demonic principalities and strongholds to destroy and lay them waste!

THE LINE OF JUDGMENT AND PLUMMET OF RIGHTEOUSNESS

Judgment also will I lay to the line, and righteousness to the plummet (Isaiah 28:17).

A plumb line has two components – the Line and the Plummet. The Line is what is stretched out over what is being "plumb lined". The Plummet is the lead, weight, or stone suspended at the end of the line. So when God sets or stretches out a plumbline, you see Him defining what the line is and what the plummet or weight is. That is why you see God saying, "Judgment also will I lay to the line." That tells you He is stretching out the ***line*** of Judgment! Then He talks about the plummet when He lays "righteousness to the plummet." The line must correspond with the plummet because they must work together. That is why in the bible you see the plummet of emptiness used with the line of confusion; the line of Samaria used with the plummet of the house of Ahab and the line of Judgment used with the plummet of righteousness. In Isaiah 28, you see God releasing the judgment of the plumbline against certain workers of iniquity.

> *Because ye have said,* ***We have made a covenant with death, and with hell are we at agreement;*** *when the overflowing scourge shall pass through, it shall not come unto us:* ***for we have made lies our refuge, and under falsehood have we hid ourselves*** (Isaiah 28:15).

These evil workers have covenants with death and hell. They have an assignment to populate death and hell with souls. They live and operate on earth but are connected to the regions of the damned. **"Her feet go down to death; her steps take hold on hell"** (Pro 5:5). They hide their activities and malicious intentions against you under the refuge and covering of lies and falsehoods. If you do not stop them, they can cause irreparable damage to your life. But with the plumbline, you will smoke them out of their hiding places and destroy them.

You see, Satan does his best work under the cover of darkness. He therefore trains his forces in the art of covert warfare to camouflage and disguise all their works and shoot arrows "privily" at the upright. They hide and embed themselves in the dark places of your life for many years waiting for the right time to strike. Years ago I prayed for a lady who was afflicted with a dangerous disease that was slowly killing her. Her whole life was turned upside down by this affliction. After I prayed for her, she called the next day to tell me that she woke up to find a dead lizard on her bed! That affliction was broken, and she regained her health.

That is what happens when your life is plumb lined! **"The strangers shall fade away, and be afraid out of their close (hiding) places"** (Psa 18:45). Once you deploy the plumbline of judgment and righteousness, strangers to your destiny who are buried deep in your life will be exposed and smoked out! Their cover will be blown, and their true identities will be revealed to you. Right now, I command all hunters, fowlers, and strangers masquerading as friends in your life to be exposed and removed in Jesus' mighty name! Now issue these plumbline decrees.

- **I loose the line of judgment and plummet of righteousness against all strangers to my destiny, to expose and destroy them in Jesus' name!**
- **I command all strangers to fade out of their hiding places in my life in Jesus' name!**
- **May the hail sweep away the refuge of lies, and may the waters overflow the hiding places of these strangers in Jesus' name!**

THE LINE OF CONFUSION AND STONES OF EMPTINESS!

STRETCHING OUT THE LINE OF "TÔHÛ" AND THE STONES OF "BÔHÛ"

> *...he shall stretch out upon it the line of confusion, and the stones of emptiness* (Isaiah 34:11).

In Isaiah 34:1-17, God judges the people of Edom with the plumbline of confusion and plummet of emptiness for how they mistreated and persecuted Israel. For this judgment God uses a special plumbline He calls the "line of confusion and the stones of emptiness". The word "confusion" in Hebrew is "tôhû" which means "to lie waste, a desolation, desert, a worthless thing, confusion, empty place, without form, vanity, wilderness". What a terrible weapon this is! And the word "Emptiness" in Hebrew is the word "bôhû" and it means "to be empty, a vacuity, an undistinguishable ruin, emptiness, void." The plumbline of confusion (tôhû - without form) and the stones or plummet of emptiness (bôhû - void) is what God used to destroy the ancient world (See Gen 1:2; Jer 4:23-27). That is what God is revealing in Isaiah 34:11.

God uses the plumbline of confusion and emptiness to destroy a land and empty it of all inhabitants and turn it into a desolate wilderness. And in the same way, when you stretch out this plumbline you will see strongholds of darkness demolished! You will see mountains of ancient demonic entrenchments uprooted and cast out! You will see their strongholds emptied of people as these souls are transferred from darkness to light! Souls that were so difficult to win will open up to the gospel as the stranglehold of darkness on their life is broken! Now set this plumbline in your region.

- **I stretch out the line of confusion over demonic principalities and strongholds ruling this region in Jesus' name!**
- **I render desolate strongholds (of gambling/witchcraft/homosexuality/etc.) in my city with stones of emptiness in Jesus' mighty name!**

THE LINE OF SAMARIA AND PLUMMET OF AHAB

And I will stretch over Jerusalem the line of Samaria, and the plummet of the house of Ahab... (2 Kings 21:13).

In Israel, there arose a King, Ahab, who angered God more than any other king Israel ever had. "But there was none like unto Ahab, which did sell himself to work wickedness in the sight of the LORD, whom Jezebel his wife stirred up" (1 Ki 21:25). Ahab and his wife Jezebel introduced Baal and Asherah worship into the land of Israel. Jezebel took this evil to the next level by executing all the prophets of God in the land. But what made God really angry with this couple was when Jezebel used Ahab's seal to plot the death of Naboth so that they could take over his vineyard. God therefore sent Elijah to tell Ahab:

> ***Behold, I will bring evil upon thee, and will take away thy posterity,*** *and will cut off from Ahab him that pisseth against the wall, and him that is shut up and left in Israel* (1 Kings 21:21).

When Ahab heard this prophecy he repented and so God suspended the judgment to the next generation. Then right after Ahab died, judgment hit his house! God raised up Jehu to destroy Ahab's household. First, Jehu killed Ahab's son Joash and executed the evil queen, Jezebel. The next day he had all seventy sons of Ahab killed. Right after that, Jehu invited all Baal worshippers to the Temple of Baal to perform a great sacrifice to that idol (See 2 Kings 10). Once that hall was filled with those idolaters Jehu had all of them killed. He then methodically executed everyone connected to Ahab. All this happened because God set a plumbline in the house of Ahab when he sent Elijah to pronounce judgment against him.

And so when God said in 2 Kings 21:13 **"And I will stretch over Jerusalem the line of Samaria, and the plummet of the house of Ahab"**, He was basically saying, *"I am bringing the same judgment I brought against Ahab's household and the city of Samaria against the city of Jerusalem!"* In other words, just as Ahab's household was destroyed, Jerusalem would be utterly destroyed. "... and I will wipe Jerusalem as a man wipeth a dish, wiping it, and turning it upside down" (2 Ki 21:13b). God allowed Babylon to come against Judah and destroy the land and kill the people because of their wickedness. Those who remained were taken captive to Babylon.

That is one of the ways of God. He can bring the same punishment He executed on a person to another person and another place in another generation many years later. As an Eternal being whatever God does (be it a reward or a judgment) never expires. "I know that, whatsoever God doeth, it shall be for ever" (Eccl 3:14). And so even now many thousands of years later you can take the plumbline God executed against Ahab's household and stretch it out against the household of wickedness in your life!

The plumbline of Ahab is especially effective against households of wickedness in covenant with Satan. God may have rescued you from such a household and now He wants to use you as a deliverer to bring judgment to that house and rescue all those being oppressed by the evil strongman ruling that house. When you deploy this plumbline you will see the strongman ruling that house dethroned and cast out and all those he oppressed set free. You will experience household salvation as all your family members and relatives get saved. At the same time, evil workers in the house will be judged and removed. That is what the line of Samaria and the plummet of the house of Ahab does. Now make these decrees with this special plumbline

- **In Jesus' name, I stretch the line of Samaria over household wickedness and household witchcraft!**
- **With the plummet of the household of Ahab, I smash and destroy all witchcraft covens operating against my life in Jesus' mighty name!**

CHAPTER 7 – The Cross!

THE CROSS

... by the Cross He triumphed over them (Colossians 2:15 WNT).

NOW I HAVE RESERVED THE best weapon for the last because the first shall be last and the last first. The Cross is the most powerful weapon of mass destruction in this war manual. I was alone in prayer early in the morning on January 18, 2018, when the LORD gave me the revelation of this weapon. He began by saying to me,

"Son, I know how much you love the revelation of the weapons of war. But do you realize that even though Moses exhausted all these weapons upon Egypt, Pharaoh still did not let the children of Israel go?

"What finally broke Pharaoh?"

"The Passover! The cross!" I exclaimed in answer to God's question

"That is right. It was the Passover that broke Pharaoh. Of all the weapons, the cross is the most powerful weapon. The cross is the weapon that ends all battles."

This revelation brought what I had been teaching on the weapons of God to a more complete and mature state. I had been teaching on the weapons of war for almost ten years when the Lord gave me this revelation because we get more revelation as we mature. As I went over all the weapons in my mind, I realized that what God was telling me was true. Even though Moses struck Egypt with several plagues Pharaoh refused to let Israel go!

> ***And Moses and Aaron did all these wonders before Pharaoh:*** *and the LORD hardened Pharaoh's heart, so that* ***he would not let the children of Israel go out of his land*** (Exodus 11:10).

And it is the same today. Many have fought several battles with the Kingdom of darkness and yet the oppression continues. Many believers, exhausted by these long drawn out battles, have given up. Many have written off spiritual warfare as a waste of time. But I want you to understand that God does not just give the enemy a knockout blow in the first minute of the warfare. He strategically increases His warfare on the kingdom of darkness until it is broken beyond repair. He softens up the enemy before He delivers the kill blow!

> ***He breaketh me with breach upon breach,*** *he runneth upon me like a giant* (Job 16:14).

Pharaoh was not broken by the first miracle Moses did when he cast his rod before him. He rather became more obstinate and doubled the burdens of Israel. Even after Moses turned the waters of Egypt into blood and struck Egypt with plagues of frogs, lice, flies, boils, hailstones, locusts, and darkness, Pharaoh still did not let Israel go because his heart had been hardened by God. The LORD God wanted to demonstrate His glory all over the earth through all these miracles He brought upon Pharaoh as Exodus 9:16 states, "And in very deed for this cause have I raised thee up, for to shew in thee my power; and **that my name may be declared throughout all the earth."** In the same way, God is manifesting His power and glory through the warfare you go through. He is training your hands for war and bringing glory to Himself at the same time. But there is a point where the warfare has to end.

THE LORD'S PASSOVER

> *And the LORD said unto Moses,* ***Yet will I bring one plague more upon Pharaoh, and upon Egypt; afterwards he will let you go hence:*** *when he shall let you go, he shall surely thrust you out hence altogether* (Exodus 11:1).

The Passover ends all battles! You see, after all the plagues Moses released on Egypt, Pharaoh still would not let Israel go. Then God tells Moses that He will bring one more plague upon Pharaoh. But this was not going to be a normal plague. This would be a plague so terrible that it would force Pharaoh to release the children of Israel. There would be no turning back, no more appeasement, no more argument, Pharaoh would be so broken "he shall surely thrust you out hence altogether." God then instructed that each household kill a lamb and anoint the doorposts and lintels of their house with the blood. And then they were to eat the roasted lamb.

> *And thus shall ye eat it; with your loins girded, your shoes on your feet, and your staff in your hand; and ye shall eat it in haste:* ***it is the LORD'S passover. For I will pass through the land of Egypt this night,*** *and will smite all the firstborn in the land of Egypt, both man and beast; and against all the gods of Egypt I will execute judgment: I am the LORD* (Exodus 12:11-12).

That night, on the 14th day of Nissan, disaster struck Egypt. The Egyptians rose in the night to see their firstborn sons dead, including Pharaoh's son (See Exo 12:29-30). Pharaoh quickly summoned Moses and commanded him to leave with all the Israelites.

> *And he called for Moses and Aaron by night, and said,* ***Rise up, and get you forth from among my people, both ye and the children of Israel; and go, serve the LORD,*** *as ye have said.* ***Also take your flocks and your herds, as ye have said, and be gone; and bless me also*** (Exodus 12:31-32).

Are you ready to breakthrough and breakout of all confinement, limitation, and oppression of the enemy in your life? You must have your loins girded, your shoes on, and your staff in your hand! It is time to go out of Egypt! It is time for you to break out of the iron furnace. God in this chapter is releasing the final blow that will conclusively end the cycle of oppression in your life. The enemies you see in your life today you will see them no more forever! After God killed the firstborn of Egypt, Pharaoh released Israel! Once you use this weapon, the enemy will let you go. But this is God's wisdom, He will have you first attack the enemy repeatedly with body blow after body blow until he is softened up for the final blow. That is why God trains you to walk in all the other weapons to confront and overcome the enemy before He finally allows you to use the cross. The best weapon is used last!

Jesus Is The Passover Lamb!

The Passover lamb is a copy and a shadow of what God did through Jesus 2000 years later. John the Baptist said of Jesus, **"Behold the Lamb of God!"** (John 1:36). Jesus was the Father's

Passover lamb! **"For even Christ our Passover is sacrificed for us"** (1 Co 5:7). That is why Jesus was so excited to celebrate the Passover with His disciples and said to them, **"With desire I have desired to eat this passover with you before I suffer"** (Luke 22:15). Jesus told the disciples that He was going to suffer. That is why the Bible calls the suffering Jesus went through, "his Passion" (see Acts 1:3). Jesus went through a very terrible ordeal during His trial and crucifixion.

Jesus appeared to me in 2005 in a dream to help me better understand what He went through. In this dream, Jesus was wearing a white robe and I was sitting close to Him on His right hand. He turned to me and we looked at each other in the face. I will never ever forget that glorious moment! I still remember seeing His golden hair parted in the middle, glowing and flowing down to His shoulders. I looked with great wonder at His beautiful face. Then He pulled up the long sleeve of His white robe to reveal His right hand all the way up to His elbow. His hand and flesh were all torn up from the Roman scourge. I gasped, horrified at how deeply mangled His hand was. I said, *"That is just how it was in Mel Gibson's film!"* Jesus said to me, **"Yes, that is how bad it was."** After that I saw Jesus being led to the crucifixion then I woke up knowing I had just been with the Lord! Oh, how wonderful the Lord is!

The crucifixion was a painful ordeal for Jesus. I know this more so because the Lord allowed me to see just a little bit of what happened. It was a very brutal death. The Prince of life had to give up his life in the most shameful and painful way possible. But Jesus considered it all joy because He and the Father knew exactly what would happen. They knew this was the only way to permanently break Satan. You see, at the beginning of His earthly ministry, Jesus cast out the devil after a 40 day fast, but Satan left Him only for a season. And now three and a half years later, Satan came against Jesus again. Concerning this Jesus said, "Hereafter I will not talk much with you: **for the prince of this world cometh, and hath nothing in me"** (John 14:30).

Jesus kept quiet to hide the real purpose of the crucifixion from the devil and his princes

---

> *Which none of the princes of this world knew: for had they known it, **they would not have crucified the Lord of glory*** (1 Corinthians 2:8).

If Satan and his princes knew that the crucifixion was going to be their undoing they would never have crucified Jesus! Never! But it was hidden from them. They did not realize that by killing Jesus they were giving Him legal access to hell. And so through His death, Jesus destroyed Satan as Hebrew 2:14-15 reveals, "... that **through death he might destroy him that had the power of death, that is, the devil;** And deliver them who through fear of death were all their lifetime subject to bondage". Through the crucifixion, Jesus was able to deliver all those who had been consigned to hell because of the fall. Through the cross, Jesus destroyed Satan and recovered all that the Devil stole from Adam. He spoiled Satan by taking back the Adamic crown and the Keys of Hell and death from him. Then Jesus led captivity captive and gave gifts to men!

> *Blotting out the handwriting of ordinances that was against us, which was contrary to us, and took it out of the way, nailing it to his cross;And having spoiled principalities and powers, he made a shew of them openly, **triumphing over them in it*** (Colossians 2:14-15). *And when he had disarmed the rulers and the authorities, he made a public spectacle of them, **triumphing over them in the cross*** (Colossians 2:15 ISV).

Jesus triumphed over the enemy with the Cross. The cross invalidates all curses, and accusations Satan has against you.

Jesus nailed all that to the cross, and with the cross, Jesus spoiled principalities and powers. He paralyzed and bound Satan and took back the spoil of souls and other things the devil had stolen from man for many generations. That is the Power of the Cross! **"For the preaching of the cross ... is the power of God"** (1 Co 1:18). The Cross is the Power of God in manifestation!

YIELD TO THE WILL OF GOD

This revelation I am sharing with you came to me in a season when I was meditating on Luke 9:23, "And he said to them all, **If any man will come after me, let him deny himself,** and take up his cross daily, and follow me" I came across this scripture in my studies and I became heavily burdened to get a deeper understanding of that verse. As I prayed during that cold winter morning on January 18, 2018, I said, *"Father what does it mean to deny oneself? I believe it has something to do with us giving up everything we have, our relationships, and our possessions to follow you. But I want to be sure, is that it? Is there something more I need to know?"* I thought it was going to take me several days or at least many hours of prayer to finally get the answer. But the Lord was very gracious. He answered me immediately. He said,

> **"Paul, to deny yourself, do what my son did at the garden of Gethsemane**
> **"You must surrender your will and take on my will!"**

You must realize how important it is to know the will of God and to do it. Jesus did not come down from heaven to do His own will but the Father's will. He said, **"My meat is to do the will of him that sent me, and to finish his work"** (John 4:34). You must seek God's will not just to know it but also to do and finish it. That was the whole pursuit of Jesus until He got to the garden of Gethsemane. For the very first time, Jesus seemed to waver. He said to His disciples, **"My soul is exceeding sorrowful, even unto death:** tarry ye here, and watch with me" (Matt 26:38). His humanity began to manifest as He pleaded with the Father saying,

---

> *O my Father, if it be possible, let this cup pass from me...* (Matthew 26:39).

Right here you see that Jesus did not want to go through with the ordeal. This was a major moment in man's history. All of heaven and earth looked on transfixed as Jesus battled with the pivotal decision. Heaven finally exhaled in great relief as Jesus said,

---

> ... ***nevertheless not as I will, but as thou wilt*** (Matthew 26:39).

Oh, what a moment in eternity! Jesus surrendered His will and took on the Father's will. It was the Father's will for Jesus to be crucified. And after another hour in prayer, Jesus said with great resolution, "O my Father, if this cup may not pass away from me, except I drink it, **thy will be done"** (Matt 26:42). Thy will be done! I want you to say this three times. Thy will be done! That is the prayer that set into motion the greatest victory of the Kingdom of heaven on earth. This is why you must give yourself to prayer. It is in prayer that you consecrate yourself to do God's will. It is in prayer that you discover God's will and receive strength to do it. **You cannot take up your cross and follow Jesus unless you first deny yourself. It is only when you surrender to God's will that you can then move on to the next step where you take up your cross.**

TAKE UP *YOUR* CROSS!

GOD GAVE YOU A CROSS TO DO WHAT JESUS DID WITH HIS CROSS!

> ***... and take up his cross daily, and follow me*** (Luke 9:23).

I asked the Father another question. I said, *"Lord what does it really mean to take up one's cross?"* He said these glorious words to me, **"When you take up your cross, you do with it whatever Jesus did with his cross.**

"Search my word diligently. Whatever you see Jesus doing with His cross, you can do with your cross. I gave you your own cross to do everything Jesus did with His cross.

"Just as Jesus spoiled principalities and powers with His cross you can do the same with your cross!"

That is when I finally understood why the Lord first began talking to me about how Moses could not break Pharaoh until he executed the LORD's Passover! Jesus talked about *you* taking up your cross in Luke 9:23 when He said, "If any man will come after me, let him deny himself, and **take up his cross daily,** and follow me." Do you see how it does not say "take up Jesus' cross"? It says "take up his cross" which means you have a cross just as Jesus had a cross. Matthew 16:24 also states,

---

> *Then said Jesus unto his disciples, If any man will come after me, let him deny himself,* ***and take up his cross,*** *and follow me.*

You are crucified with Jesus because you also have a cross as Paul said, **"I am crucified with Christ"** (Gal 2:20). It is with this cross that you die daily to the flesh, the world, and the devil. And with your cross, you can also spoil principalities and powers just as Jesus did with His cross! That is what it means to be crucified with Jesus! With the cross, you can conclusively end Satan's domain in your life. **With your cross, you can spoil principalities and powers and make a show of them openly and triumph over them!** With the cross, you can bind and spoil the strong man ruling your life, your city, your state, and your nation. That is the power of the cross. It is a great weapon of mass destruction! Now make this declaration.

Everything Jesus did with His cross, I can do with my cross! Today I take up my cross, and I do with it everything Jesus did with His cross!

The Cross is the weapon of war that ends battles. With the Cross, God "**... maketh wars to cease unto the end of the earth**" (Psa 46:9). This is one of the glorious works of our Father. God wants to do with you what He did with Moses and with His Son Jesus Christ on Passover. When you take up your cross, God will personally back your words and end the seemingly endless warfare in your life. He will cause all wars in your life to cease! And when you break evil strongmen with the cross, they will beg you to leave their dungeons and prisons. They will be glad to see you go because God Himself will fight them for you when you unleash the power of Passover, that is, the power of the Cross, the weapon that ends all battles! Do that right now with these decrees:

- **I invalidate and nail to the cross every curse operating in my life in Jesus' name!**
- **By the power of the cross, I remove every obstruction to my destiny in Jesus name!**
- **By the power of the cross, I tear up every veil on my life in Jesus' name!**
- **By the power of the cross, I confront the strongman terrorizing, oppressing, and hindering my life, in Jesus' name!**
- **By the power of the cross, I spoil the house of the stranger and I recover all that has been stolen from my life in Jesus' name!**
- **By the power of the cross, I make a show of every principality and power operating in my life and expose their works, in Jesus' name!**
- **By the cross, I overcome and triumph gloriously over every principality and power operating against my life, in Jesus name!**
- **I loose the power of the cross against every dark prince and strongman ruling my city and I spoil their strongholds in Jesus' name!**

- **By the power of the cross, I take captivity captive and translate the souls of this city from hell and death into the Kingdom of God in Jesus' name!**

Now do not just be a hearer of the word – be a doer! Take these seven weapons you just learned and put them to work immediately and may Jehovah, the Man of War, and all of heaven back you! And with that, I must end.

Your friend and brother in Christ,
Paul K. Adom

ENDNOTES

INTRODUCTION

1. Encyclopedia Britannica, *"Weapon of mass destruction (WMD)" https://www.britannica.com/technology/weapon-of-mass-destruction* (Accessed July 11, 2020)
2. David E Taylor, *The Empire of God Part 1* (Florissant, MO: Joshua Media Ministries International, 2018) p29
3. James Strong, *The New Strong's Expanded Exhaustive Concordance of the Bible* (Nashville, TN: Thomas Nelson Publication, 2010) "hoplon" G3696 p276

CHAPTER 1 – THE MORNING STAR

1. European Southern Observatory, *"Venus' orbit and visibility - day, night and seasons on Venus" https://www.eso.org/public/outreach/eduoff/vt-2004/Background/Infol2/EIS-D3.html#:~:text=The%20orbit%20of%20Venus%20which,of%20the%20Sun%20and%20Moon.* (Accessed July 11, 2020)
2. F. Brown, S. Driver, C. Briggs, *The Brown-Driver-Briggs Hebrew and English Lexicon* (Peabody, Massachusetts: Hendrickson Publishers, Inc, 2007) "kûn" H3559 BDB p465c

3. James Strong, *The New Strong's Expanded Exhaustive Concordance of the Bible* (Nashville, TN: Thomas Nelson Publication, 2010) "Shachar" H7837 p276
4. IBID, "bôqer" H1242 p44
5. IBID "mishchâr" H4891 p174
6. IBID "Shachar" H7837 p276
7. Albert Barnes' Notes on the Whole Bible, Job 38:14 Commentary, *https://www.truthaccordingtoscripture.com/commentaries/bnb/job-38.php#.XupTLZpKj3g* (Accessed June 17, 2020)

Chapter 2 – The Sword of The LORD

1. James Strong, *The New Strong's Expanded Exhaustive Concordance of the Bible* (Nashville TN, Thomas Nelson Publication, 2010) yâ'ad H3259 p116
2. F. Brown, S. Driver, C. Briggs, *The Brown-Driver-Briggs Hebrew and English Lexicon* (Peabody Massachusetts, Hendrickson Publishers, Inc, 2007) yâ'ad BDB p416d
3. Spiros Zodhiates, *Word Study Series: The Complete Word Study Dictionary* (USA, AMG International, Inc Revised Edition 1993) hoplon 3698 p1051

Chapter 3 – The Arrow of God

1. James Strong, *The New Strong's Expanded Exhaustive Concordance of the Bible* (Nashville, TN: Thomas Nelson Publication, 2010) H2671, H2686, H2673 p95
2. BBC NEWS, *Russia deploys Avangard hypersonic missile system* (ref *https://www.bbc.com/news/world-europe-50927648* (Accessed June 18, 2020)
3. Matthew Henry's Commentary on the Whole Bible, *Psalm 120:4 https://www.christianity.com/bible/commentary.php?com=mh&b=19&c=120* (Accessed June 18, 2020)

Chapter 4 – The Iron Rod

1. James Strong, *The New Strong's Expanded Exhaustive Concordance of the Bible* (Nashville, TN: Thomas Nelson Publication, 2010) mâshal H4910, p95

Chapter 5 – The Cup of the LORD's Fury

1. F. Brown, S. Driver, C. Briggs, *The Brown-Driver-Briggs Hebrew and English Lexicon* (Peabody, Massachusetts: Hendrickson Publishers, Inc, 2007) zal'âphâh / zil'âphâh H2152 BDB p273c

Chapter 6 – The Line and Plummet

1. Encyclopedia Britannica, *"Plumb Line" https://www.britannica.com/technology/plumb-line* (Accessed June 6, 2020)
2. James Strong, *The New Strong's Expanded Exhaustive Concordance of the Bible* (Nashville TN, Thomas Nelson Publication, 2010) tôhû H8414 p295
3. IBID, bôhû H1242 p34

Made in the USA
Monee, IL
21 December 2021

86758915R00059